SpringerBriefs in Applied Sciences and Technology

Continuum Mechanics

Series editors

Holm Altenbach, Magdeburg, Germany
Andreas Öchsner, Southport Queensland, Australia

These SpringerBriefs publish concise summaries of cutting-edge research and practical applications on any subject of Continuum Mechanics and Generalized Continua, including the theory of elasticity, heat conduction, thermodynamics, electromagnetic continua, as well as applied mathematics.

SpringerBriefs in Continuum Mechanics are devoted to the publication of fundamentals and applications, presenting concise summaries of cutting-edge research and practical applications across a wide spectrum of fields. Featuring compact volumes of 50 to 125 pages, the series covers a range of content from professional to academic.

More information about this series at http://www.springer.com/series/10528

Frank A. Coutelieris · Antonios Kanavouras

Contributing authors: Kostas Theologou, Spyridon Stelios

Experimentation Methodology for Engineers

 Springer

Authors
Frank A. Coutelieris
Department of Environmental
   and Natural Resources Management,
   School of Engineering
University of Patras
Agrinio
Greece

Antonios Kanavouras
Department of Food Science
   and Human Nutrition
Agricultural University of Athens
Athens
Greece

Contributing authors:
Kostas Theologou
Department of Humanities, Social Sciences
   and Law
National Technical University of Athens
Athens
Greece

Spyridon Stelios
Department of Humanities, Social Sciences
   and Law
National Technical University of Athens
Athens
Greece

ISSN 2191-530X   ISSN 2191-5318   (electronic)
SpringerBriefs in Applied Sciences and Technology
ISBN 978-3-319-72190-3   ISBN 978-3-319-72191-0   (eBook)
https://doi.org/10.1007/978-3-319-72191-0

Library of Congress Control Number: 2017962022

Printed on acid-free paper

This Springer imprint is published by the registered company Springer International Publishing AG part
of Springer Nature
The registered company address is: Gewerbestrasse 11, 6330 Cham, Switzerland

The original version of the book was revised: Corresponding authors' affiliations have been changed. The erratum to the book is available at https://doi.org10.1007/978-3-319-72191-0_6

Contents

Introduction 1
Frank A. Coutelieris and Antonios Kanavouras

Unveiling Scientific Knowledge for an Engineering Model 5
Kostas Theologou, Spyridon Stelios and Antonios Kanavouras
  A. The Constituents of Knowledge 5
  B. Science and Scientific Knowledge 9
  The Emergence of the Philosophy of Science: From Kant to Popper
  and Kuhn 11
    Immanuel Kant (1724–1804) 11
    Karl Popper (1902–1994) 14
    Thomas S. Kuhn (1922–1996) 15
    Imre Lakatos (1922–1974) 17
    Paul Feyerabend (1924–1994) 18
    B1. Theory 19
    B2. Experiment 20
    B3. Scientific Truth and Social Construction 22
  C. Modeling and Scientific Representation 24
  Representation 25
  D. The Case of Engineering 26
  References 27

Scientific Research—Perspective, Awareness and Criticism 31
Antonios Kanavouras, Frank A. Coutelieris, Kostas Theologou
and Spyridon Stelios
  Introduction 31
  A. The World of Phenomena 34
  Research as a Concept 36
  The Human Factor 36
  Interpretations and Explanations 39

Realizing Physical Phenomena 40
The Essentials for Understanding 41
The Cycle of Understanding 44
B. Research in Practice 49
Creating and Reproducing Phenomena 50
Experimentation 51
Literature Search 51
The Language of Science; the Terms of Truth 52
Information 54
Data 56
Literature Briefs 58
Engineering of the Experimentation 59
Research Planning 60
Experimental Design 64
Experimental Strategy and Experimental Results 65
Engineering Based Design 66
Lab 66
Equipment 69
Tools 70
Efficiency, Effectiveness, Economy 71
Results 71
Engineering 72
C. Mathematics 73
Mathematical Modeling 74
Speculations, Calculations, Models and Approximations 75
References 77

**On the Development of Engineering Assets—The MATRIX
Scheme** 81
Frank A. Coutelieris and Antonios Kanavouras
Introduction 82
Similarity 84
Theory 85
Knowledge Classification Scheme 87
Decisive Rules 90
The Mathematics of Matrix 91
Applications of the MATRIX Scheme on Engineering Problems 94
Application I: Oxidation of Packaged Olive Oil 95
Application II: Adsorption in Granular Media 100
Mathematical Proofs 102
References 103

Conclusions 107
Antonios Kanavouras, Frank A. Coutelieris, Kostas Theologou
and Spyridon Stelios
    Closing Note 110

Erratum to: Experimentation Methodology for Engineers E1

Index 113

Marketing Note

The contribution of the book concerns its comprehensive discussion of the main topics related to the initiation of an experimentation process and the validation of its outcome via the inclusion of state-of-the-art applications throughout a strict, mathematically defined, manner. In addition, this book presents novel theoretical, mathematical and experimental developments, providing a self-contained major reference aiming at being appealing to both scientists and engineers. Furthermore, these topics are encountered in a variety of scientific and engineering disciplines.

The book initially presents a solid theoretical approach based on historical models of thinking about science. Then, via an epistemological approach, the consequent practical implications on the formation of a hypothesis, the considerations of the existing knowledge and its classification and validation, are presented for assisting the experimentation interest, though a solid engineering-based approach. Accordingly, the transition from the knowledge classes to the experimentation parameters according to the contributors for the phenomena evolution and the properties of the systemic descriptors established within this book are also discussed. This discussion leads into identifying the major experimenting requirements which may allow the researchers to focus on those of the conditions that when applied during the experiment may or may not satisfy, a potential disclaim of the initial hypothesis. The consequent deriving experimental outcome has also been mathematically validated for the similarities among the existing knowledge and any knowledge obtained through a research process. The complete methodology presented in this book offers a justified tool towards the minimization of research effort wastes, as far as it can identify the lack(s) of knowledge, thus the areas of interest where the current research should work on. The special features of this book are (a) the use of state-of-the-art techniques for the classification of knowledge; (b) the consideration of a realistic systemic world of engineering approached phenomena; (c) the application of severe mathematical techniques for identifying, describing and testing the similarities in the research results and conclusions; and (d) the experimental investigation of relevant phenomena.

To conclude, this book delivers a methodological approach on the experimentation and/or simulation processes for scientist dealing with the potentiality of disclaiming a hypothesis regarding a physical phenomenon, up to the validation of the research results and the design of upcoming experiments.

Introduction

Frank A. Coutelieris and Antonios Kanavouras

In the times that science and technology have determined their existence as such, a numerous attempts to rejuvenate the various scientific fields either from within or via importation from other fields, there was a need to revitalize the cognition about science, that should doubt and turn upside down not just the established knowledge borders but the cornerstones of the scientific paradigms, and thus the prominence of science itself. It is indeed well known that the more radical the idea, the more difficult it is to be accepted and, in many cases, these changes were not easily and smoothly accepted and adopted by those believing in the eternal and absolute picture of present (modern) science.

Currently, it is usually the case that scientific problems and issues are viewed through both politically and ethically structured points supported by mono-dimensional theories, believes, technological formalism(s), truths and pure intentions. History has indicated quite clearly that there is not a single scientific theory which has been extensively incorporating the reality and captured any object within the scientific paradigms. On the contrary, they remain open, uncompleted, inadequate and full of gaps, uncertainties and unknowns. Yet, researchers dealing with those gaps force the progress through their attempts to fill them by re-working meta-theories, in an endless cycle of understanding, progressing, criticizing, repositioning and so on. Within this process, an error may lead the system to further complexity for resolving it or may lead it to simply fail its functionality. This uncertainty may drive engineers to serious confidence issues. A potential antidote may be an efficient establishment of distinctions between truth and error for both highly complex as well as for lower complexity systems. Having said that, we may quite safely conclude that the way forward would be a potential distinction between science and pseudo-science. In other words, science is what may reveal all that is to be revealed though research and consequent experiences, versus pseudo-science which is an illusionary, shelf-constructed outcome of unjustified and restricted conclusive remarks. Therefore, pseudo-science is highly vulnerable to criticism and pure logic. In that sense, the scientific ways through which science is functioning and progressing affect its core body and mind, far beyond the contradiction among

© The Author(s) 2018

F. A. Coutelieris and A. Kanavouras, *Experimentation Methodology for Engineers*, SpringerBriefs in Continuum Mechanics, https://doi.org/10.1007/978-3-319-72191-0_1

empiricism and pragmatic idealism. Having said that, it goes without much doubt, that the inner and deeper key-problem of science is thus, revealed along with the problem of any knowledge. That is in fact, the relationship between the observer-subject and the under-observation object. Could it then be that in more than one field and in more than one occasion, the object under observation has been constructed by the subject-observer via a cognitional process of description?

This book has been originated from an ambitious attempt to investigate an efficient way for the research on physical phenomena to progress and consequently, how the relevant experimentation should be applied for maximizing the benefits towards that. In such a quest for improving, the essential first step should be the challenge for knowing and understanding of what is currently available. Furthermore, and beyond the availability of any form of knowledge, we needed to understand what has indeed science revealed to us from all that is worth revealing and what has been still hiding. How far have we been in revealing the complete picture of the physical world in terms of its inherent phenomena and how much of this may be applied in a constructive and efficient way, i.e. within an engineered conceptualized manner for the engineering oriented fields. At this point, we have decided to proceed by taking into consideration the "motives" of the physical phenomena, meaning that we were to deal with the applied physical laws that guide the developments in a system, rather than the "causes" or "reasons" for something to happen. These later ones are the material for the physical phenomena to develop, but do not necessarily define their existence. The conditions, the "atmosphere" under which the phenomena will develop, is what drives the evolution in any physical system and that has been defined as the main "knowledge" point of interest.

But since experience is guiding our perceptions of the world and human subjectivity defines their perception, how do we know what we know? And how do we know what we do not know? How do we form a scientific hypothesis, what influences our decision, which way do we choose to approach it, which elements do we really consider and value the most for creating new experiences? What do we get out of any experience of ours and would that have been the same if that would have been others' experience? Working up to scientific expectations means working on a more objective way. That asks for a solid background where pure knowledge itself, rather than the subject that knows, has the primary role. Where the understanding is intra-subjective and the individual experience is valid only via the impact it provides to the holistic knowing process, we here in called the "cycle of understanding". Then, consciousness becomes a critical factor for the swing from nature as an epiphenomenon and nature as epicenter, the dialectic of interactions and intervention in science. The issues of maturity may define the progress in the endless road of knowledge.

We do realize that considering science and scientific evidence as "subjective", meaning that experimentation and furthermore the use of research outcome is depended on individuals, rather than a globally accepted asset, has created quite an uncomfortable situation, to say the least, within the academic communities since the '60s. That view was a major driving force for this work. We do consider that efficiency in scientific research should not be depended on who is performing the research, or what type of equipment was used, at which part of the world or under which conditions of working. We have aimed in supporting a uniform way of approaching the physical phenomena, a repeatable and reproducible way of feeding back to the existing knowledge via experience that reduces similarity and enhances innovation and entrepreneurship in the lab. The way forward for the authors has been to create a handy, yet well reasonable, "tool" for approaching the existing knowledge and be justifiably capable of identifying the knowledge gaps and potential open space for investigations. It is authors' solid opinion that research topics' hypothesis, research objectives and the ways of working, should be based on pure logical reasoning.

This approach has been supported throughout this book via theoretical and practical considerations. The authors tried to understand the cognitional way of working against a hypothesis, therefore, their approach was concretely built on clear empirically derived evidences regarding the phenomena involved, as well as practical discriminations allowing us to further re-construct the systemic phenomena, utilizing solid considerations on properties' relationships and cohesive inter-systemic views.

This book, after the introductive part, initially presents a solid theoretical background regarding the knowledge itself as well as several relative topics, such as research, science, etc. (Chapter "Unveiling Scientific Knowledge for an Engineering Model"). In this fundamental chapter, we had the decisive contribution of Kostas Theologou and Spyridon Stelios, which actually runs across the whole book. After establishing the necessary terminology and glossary, the next part (Chapter "Scientific Research—Perspective, Awareness and Criticism") deals with the idea of "studying a phenomenon", i.e. with the "circle of understanding", to identify a sufficient experimentation technology. Again, in this chapter we also had the contribution by Kostas Theologou and Spyridon Stelios. A mathematical description and validation follows (Chapter "On the Development of Engineering Assets—The MATRIX Scheme"), where the mathematical proofs are accompanying by applications of engineering interest. Furthermore, a conclusive section (Chapter "Conclusions") summarizes the main points of all the above and all four contributing authors tried to depict comprehensively our ideas in this monograph. Finally, Kostas Theologou prepared the index of all the significant ideas, entities and names included in this book.

Unveiling Scientific Knowledge
for an Engineering Model

Kostas Theologou, Spyridon Stelios and Antonios Kanavouras

For, after all, what is man in nature? A nothing in comparison with the infinite, an absolute in comparison with nothing, a central point between nothing and all. Infinitely far from understanding these extremes, the end of things and their beginning are hopelessly hidden from him in an impenetrable secret.

Pensées 1660
Blaise Pascal (1623–1662)

A. The Constituents of Knowledge

The Greek word for science is "episteme" (ἐπιστήμη) which derives from the verb ἐπίσταμαι (*epistamai, I know very well*). So, science is not merely knowing but knowing well. Knowledge is the ultimate goal of all scientific inquiry and this explains why all scientific research should state beforehand its significance and value.

The branch of philosophy that attempts to formulate normative criteria for what should be considered as knowledge is *Epistemology*. Traditionally, philosophers distinguish three kinds of knowledge: (a) theoretical knowledge (Logos) ('know that'), (b) practical knowledge ('know how'), and (c) knowledge by acquaintance ('know him/her/it'). One of the problems of *Epistemology* is whether any of these kinds is more fundamental or *basic* than others.

The Platonic dialogue *Theaetetus* 'inherited' to the philosophers and scientists the classical definition of knowledge (ἐπιστήμη, *scientia*), as *a true and justified belief* (Plato, *Theaetetus*, 201c–d). This traditional and *tripartite* approach to *knowledge* (*scientia*) claims that *justified true belief* is necessary and sufficient condition for acquiring knowledge. A rational agent S knows that *p* **if** *p* is true, S believes that *p* and S is justified in believing that *p* (Ichikawa and Steup 2017).

Of course, the classical definition has been questioned as to the universality of its application. The two most notable cases were presented by American analytic philosophers Gettier (1927–) and Robert Nozick (1938–2002). Gettier (1963) put

© The Author(s) 2018
F. A. Coutelieris and A. Kanavouras, *Experimentation Methodology for Engineers*, SpringerBriefs in Continuum Mechanics,
https://doi.org/10.1007/978-3-319-72191-0_2

5

forward examples of justified true belief that one wouldn't eagerly define as *knowledge*. Gettier's counterexamples mind the following general form:

- An agent S justifiably believes that proposition p is true.
- However, p is false. S correctly infers that if p is true, then q is true.
- So, S justifiably believes q.
- It turns out that q is true, but not because of p.
- Therefore, S has a justified true belief that q.

In Gettier-concept examples, the error of the case is generated by assuming an *inference* based on a *false* premise (namely, p), though there is *sufficient evidence* to believe that premise is true. Nozick's definition (1981) is an alternative to the classical description of 'knowing'. It uses the concept of *tracking the truth* to give priority to sound causal connection between facts and beliefs in a direct way, i.e., without the intervention of the justification 'catalyst'. A tracking theory of knowledge is one that describes knowledge as a belief that tracks the truth in a reliable way. It is an attempt to deal with Gettier counterexamples.

Nozick describes four conditions for how a person, S, can have some knowledge of a proposition p. Specifically, in order for S to know p:

(i)   proposition p is true,
(ii) S believes that p,
(iii) if it were not the case that p (i.e., if p), S would not believe that p, and
(iv) if it were the case that p, S would believe that p.

	M1 recommends	M2 recommends	Does the person believe p or believe *not-p*?
Case I	Believe p	Believe p	Believes p
Case II	Believe p	Believe *not-p*	?
Case III	Believe *not-p*	Believe p	?

Nozick (1981)

The main outcome deriving through the formulation of his hypothetical method is the need for an accurate reference to alternative conditions that are compatible with the *hypothetical negation* of the fact (represented by proposition p). In the end, knowledge is such because it tracks the truth. *Justification of a belief is only valid insofar as it reliably keeps track of what is true.*

Turning back to the classical definition, the correlation φ of concepts Truth (T), Justification (J) and Belief (B) [$K = \varphi(T, J, B)$] implies, in principle, a choice of beliefs. Those that are true are being chosen but are also accompanied by evidential elements that support them, i.e., the ones that are justified are the ones chosen. This option incorporates also a very general approach to knowledge: starting from a certain belief, gradually and through a process of controlling the conditions in which this belief is expected to be true, certain elements (or reasons) arise that either reinforce or challenge it in an objective way. If it happens to be "sufficiently" strengthen, then we are dealing with a justified belief (..., J, B) and thus we have approached knowledge sufficiently and where possible. However, from this process, the condition of truth (T) is absent. And it is truth that must be objectively (irrespectively of our own awareness) satisfied, in order for a belief to correspond to knowledge (Koutoungos 2012).

Function φ represents a normative scheme of selecting beliefs that is compatible with the process of searching for knowledge. This rule of selection and process of searching have a kind of naturalness in their manifestation: what beliefs do us have to choose or how justified they ought to be for us in order to get to the truth? Let us look in more detail each set:

A1. Belief Belief is a necessary condition of knowledge. This means that the epistemic relationship φ is manifested in a 'belief environment.' Belief is generally what someone takes for granted. That is, the person who believes in something is bound to it as it is. Forming beliefs is one of the most basic and important features of the mind. Contemporary analytic philosophers generally use the term "belief" to refer to the attitude we have, roughly, whenever we take something to be the case or regard it as true. "To believe something, in this sense, needn't involve actively reflecting on it: Of the vast number of things ordinary adults believe, only a few can be at the fore of the mind at any single time. Nor does the term "belief", in standard philosophical usage, imply any uncertainty or any extended reflection about the matter in question" (Schwitzgebel 2015). This does not mean, of course, that we stop believing in something if we happen to have something else in our mind. Beliefs are not static and by no means isolated episodes within time. This non-static nature of beliefs has given rise to a research field of formal epistemology that investigates belief revision. Generally, research in this field is divided into normative (probabilistic or non-probabilistic) and descriptive methodological approaches.[1]

A2. Justification Obviously, within this epistemological process, our actual choice is set in *justified* beliefs. Simply put, justification refers to whether specific beliefs are held in a responsible manner. It cannot be extended to the truth condition whose presence in the classical definition seems to depend on other more restricted choices. There exists already a problem of genuinely distinguishing justified belief from true belief because justified beliefs are more likely to be true. Set J contains those circumstances

---

[1]The *Bayesian* update of beliefs is the most known probabilistic approach, whereas *Belief Revision* as suggested by Peter Gärdenfors and fellow researchers is the most widespread non-probabilistic one (see Gärdenfors 1992). With regard to the descriptive approach, belief revision is studied mainly in the areas of cognitive and clinical psychology, communication, etc. (see also Stelios 2016).

allowing for an amount of information/evidence to be linked with beliefs. Thereby these instances are testimony (or justification) of corresponding beliefs. Justification is often used synonymously with rationality, but could also be understood in a normative context. That is, having justified beliefs is better, in some sense, than having unjustified beliefs and determining whether a justified belief indicates whether we should, should not, or may believe, a particular proposition. Finally, there is also a rather more naturalistic (or science-based) approach that considers the relationship between belief-forming mechanisms and objective reality.

A3. Truth

Apparently the 'belief environment' is not a sufficient premise of the classical epistemic relationship. Satisfying the need of covering the gap between the subjectivity of set B and infallibility of (objective) knowledge, truth is added along with its pure objectivity. As far as set T is concerned, the permissible use of descriptions, such as *true*, *truth* or *be true* seems to depend on rules that operate at more than one levels, the most dominant of which is our relationship with the empirical or external world.

Within this framework, the Coherence and Correspondence theory, with their variants, are the two basic theories of truth. The theory of Coherence states that the truth of any (true) proposition consists in its coherence with a certain set of propositions. It differs from its main competitor, the Correspondence theory, in two key positions: (a) regarding the relation of the propositions with their truth ('proposition' is considered here as a bearer of a truth value, whatever it may be), where in Coherence, this relationship is coherence, and in Correspondence it is correspondence, and (b) regarding the interpretation of the conditions of truth, where in Coherence the prerequisites for the truth of propositions are other propositions, while in Correspondence theory the conditions of truth are rather objective characteristics of the world (without excluding, of course, the use of propositions as their truth values).

Although quite opposed to each other, both theories offer a meaningful perception of truth. Contrary to pragmatist theories, they argue that truth is a property of propositions that could be analyzed in relation to the types of conditions they should satisfy, as well as to the relationship between propositions under these circumstances. In contrast, for a pragmatist, in general, truth does not have any particular nature beyond what is contained in simple claims. For example, the proposition "this liquid is green" is true only if the liquid is actually green. In this respect, those who seek the real nature of truth will always be disappointed since they are looking for something that does not exist.

B. Science and Scientific Knowledge

Epistemology is the study of knowledge and how this is acquired. Science is (logical) knowledge obtained through inference of facts determined by calculated experiments. Through the way that science itself has been developed, it is scientific knowledge that now includes a broader usage especially in social sciences where it is discussed as meta-epistemology, or generic epistemology, and is, to some extent, related to a cognitive development.

Science along with the nature of scientific knowledge has also become a field of interest for Philosophy. Philosophy of Science is concerned with the assumptions, foundations, and implications of Science and as a discipline is rather defined by an interest in one of a set of *traditional* problems in its foundational concerns. Many philosophers of science pose or ask central questions in science, and they consider these questions as applying to particular sciences (e.g., philosophy of Mathematics, Biology or Physics). In a similar mode, philosophers of science also use contemporary scientific results to draw philosophical morals. Most practitioners in Philosophy of Science are not only philosophers. Several prominent scientists have effectively contributed to the field and still do. Overall, the main task of the Philosophers of Science is to produce the criteria on to what qualifies as Science.

As with most aspects of science, knowledge obtained through experience or empirical evidence is a posteriori, an afterwards knowledge. The pure existence of a term like a posteriori implies that there is also a counterpart involved. In this case, that is the a priori knowledge, meaning previous to. For the knowledge to exist prior to any experience, it requires that there are certain *assumptions* one takes for granted. These assumptions refer to necessary conclusions arising from specific logical premises that precede empirical observations. Examples of a priori knowledge are mathematics and analytic propositions. As it will be presented in the next section, the term *'analytic proposition'* was introduced by Immanuel Kant at the beginning of his *Critique of Pure Reason* (1781) and refers to any statement that is true based on proof by contradiction, i.e., its denial leads to contradiction (Fig. 1).

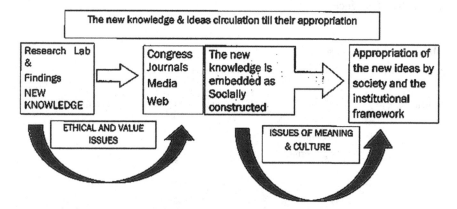

Fig. 1 New knowledge (Theologou 2014)

Early modern theories of knowledge, especially those advancing the influential empiricism of John Locke (1632–1704), were based implicitly or explicitly, on a model of the mind which associated *ideas* to *words*. This analogy between *language* and *thought* laid the foundation for a graphic conception of knowledge in which the mind was treated as a *tabula*, a container of content that had to be stocked with facts reduced to letters, numbers or symbols. This led to a situation in which the *spatial alignment of words on the page carried great cognitive weight*. That was so much so that educators paid a very close attention to the visual structure of information on the page and in notebooks.

Now, it could be argued that knowledge is 'transformed' into science when its three constituent parts, as mentioned above, receive a more rigorous methodological form. The Table 1 describes this relationship and is the basis of our analysis.

Before moving any forward, we believe it is necessary to make a reference to the relationship between scientific knowledge and technology. *Scientific knowledge* tends to be normative and consists of an articulated grid of observations, theories and laws. On the technological level, the elements of knowledge that correspond to the above scientific components are accordingly, actions, theories and rules. In addition, scientific laws try to describe, in an objective (not influenced that is, by particular perspectives or personal/collective interests) manner, the standards of the experienced *natural* phenomena or events and are *considered* to a great or lesser extend true, whereas technological rules predetermine the ways of acting and they are characterized to a great or lesser extend as efficient.

But what exactly is it that the *natural* scientist claims to know about the *natural world* and how such claims are being assessed? To either determine ultimate truth and/or expose questions about nature that are not yet answered, science is said to consist of two working areas, *theory* and *experiment*. Theories struggle to provide an account on how the world stands, while experiments and technology try to change the world, *to adjust it to the new theory*, no? Common understanding of science states that theories are, or should be, *systematic, coherent, predictive* and *broadly applicable*. But things are not always so simple! The foundations, methods, and implications of science cover a wide range of study in what has been termed as Philosophy of Science.

Table 1 Knowledge and science

A. Knowledge	1. Belief	2. Justification	3. Truth
	↓	↓	↓
B. Science	1. Theory	2. Experimentation	3. Scientific Truth
	↓		
	Modeling		
	Engineering		

The Emergence of the Philosophy of Science: From Kant to Popper and Kuhn

Immanuel Kant (1724–1804)

As shown so far, Philosophy of Science is indissolubly linked to the theory of (acquiring) knowledge. Despite its diffusion into different research disciplines and topics, it does not cease to focus primarily on the way (or ways) in which scientific knowledge is actually true knowledge. In this context, Immanuel Kant's contribution has been a major one. Through his work, he redefined in a strict conceptual manner the relationship between experience and the outside world and set the limits of any possible human knowledge.

In his monumental work *Critique of Pure Reason* (1781) Kant described the theory of *transcendental idealism*. This kind of idealism was not equivalent to mind-dependence in the sense of Berkeley's idealism. It made reference to a reality which is beyond our grasp, in *things in themselves (Dinge an sich selbst)*.[2] Through the lens of *transcendental idealism* appearances of things are regarded as "mere representations and not as things in themselves, and accordingly that space and time are only sensible forms of our intuition, but not determinations given for themselves or conditions of objects as things in themselves. To this idealism transcendental realism is opposed as it regards space and time as something given by themselves (independent of our sensibility)" (Kant 1781/1998). Generally, his theory combined two rival and often conflicting philosophical schools: *idealism* and *realism*. As far as idealism is concerned, Kant's reality is our own creation, and in terms of realism, the cause of every experienced phenomenon is the reality that exists independently of us.

One of his greatest contributions was the distinction between a priori and a posteriori knowledge. A priori cognitions are "universal cognitions, which at the same time have the character of inner necessity, must be clear and certain for themselves, independently of experience; hence one calls them a priori cognitions: whereas that which is merely borrowed from experience is, as it was put by Kant, cognized only a posteriori, or empirically" (Kant 1781/1998). Kant sought to defend against empiricists the claim of the possibility of universal and necessary knowledge—a knowledge originating independently of experience. In human cognition there are actually pure, a priori judgments. An example taken from sciences concerns all the propositions of mathematics. By contrast, a posteriori knowledge or knowledge derived from any particular experience is dependent on experience or empirical evidence, as with most aspects of science and personalized knowledge.

---

[2]A "thing in itself" is one half of the appearance/thing in itself distinction, which Kant originally defined at A491/B519 in terms of their existence: appearances have no existence "grounded in themselves" while things in themselves do (see Stang 2016).

In conjunction with the a priori/a posteriori distinction, Kant advocated a blend of rationalist and empiricist theories. Despite the existence of an a priori knowledge, humans cannot know anything about any real thing in the world (*"thing in itself"*) or about what is not an actual or possible object of experience. In this way, as Russell (1912) noted, Kant tried to reconcile and harmonize the contentions of the rationalists with the arguments of the empiricists (see Russell 1912/2001). According to Kant, an a priori cognition is transcendental, or based on the form of all possible experience, while an a posteriori cognition is empirical, based on the content of experience. Although all of our cognition begins with experience, it does not follow that the former arises from the latter. Thus, Kant formulated a *scientific metaphysics* of his own, establishing that what can be known as a priori is also limited to that which is required for ordinary experience and its extension into natural science.

In the introduction to the *Critique*, Kant argues that our mathematical and physical knowledge of nature requires certain judgments that are (i) a priori and (ii) *synthetic* rather than *analytic*. Kant conceived an analytic statement as one that attributes to its subject no more than is already conceptually contained in it. Within this conceptual frame, the predicate is obtained by merely analyzing the subject.[3] Thus, a *synthetic* statement is the one that goes beyond this conceptual containment.

For Kant, space and time are a priori intuitions. It is not possible to meaningfully conceive of an object that exists outside of time and has no spatial components. The unity of time implies the unity of the self and the determinate position of all objects. The unity of space depends on the unity of the subject and on the ability of the self to assign representations of objects determinate positions in space (see Kant 1781/ 1998). What gives unity to the different representations in any judgment is the concept of understanding (*Verstand*). Understanding is explained as a faculty for thinking, a faculty of concepts or judgments and finally as the faculty of rules. It is distinguished from Reason (*Vernunft*) which is the faculty of principles (Kant 1781/ 1998, A126). Understanding and Reason can only be applied to things as they appear phenomenally to us in experience. What things are by themselves, meaning independent of our cognition, remains limited by what is known through phenomenal experience.

Following Aristotle, Kant refers to the pure concepts of the understanding that pertain to objects a priori. These concepts or *categories* are necessary to achieve order among the intuitions of objects presented in the form of time and space. Kant distinguishes 12 categories (Kant 1781/1998, B106):

(1) Of Quantity: Unity, Plurality, Totality,
(2) Of Quality: Reality, Negation, Limitation,

---

[3]According to Quine (1951) Kant's conception should be restated as "a statement is analytic when it is true by virtue of meanings and independently of fact". Based on that, a priori *analytic* statements are thought to be true in virtue of their meaning, while a posteriori *analytic* propositions are thought to be true in virtue of both their meaning and certain facts about the world.

(3) Of Relation: Of Inherence and Subsistence (*substantia et accidens*), Of Causality and Dependence (cause and effect), Of Community (reciprocity between agent and patient),

(4) Of Modality: Possibility—Impossibility, Existence—Non-existence, Necessity —Contingency.

Categories are entirely different from the appearances of any object and in order to relate to specific phenomena, they must be "applied" through *schemata*. Because categories are, according to Kant, heterogeneous with sense intuition, these schemata provide the necessary link with the sensed phenomenal appearances. So, for every one of the 12 a priori concepts of the understanding there is a connection with a phenomenal a posteriori appearance.

Categories have a determinate cognitive use only when applied to spatiotemporal data and to appearances ("*phenomena*"). Therefore, by means of the categories, things as they are in themselves ("*nooumena*") might *be thought* but not known. Even though we can understand that things are different from what they appear to be, any knowledge of things by themselves, beyond experience and when achieved solely through pure reason, is totally inaccurate. For Kant, the only object for the human knowledge is experience. Categories give us knowledge only of things as they appear with sensibility ("*phenomena*"). So, *phenomena* depend upon the conditions of sensibility, space and time, and on the synthesizing activity of the mind. Although through pure understanding (*nous*) we may think of objects independently of their being given in sensibility, we can never cognize them as non-sensible entities ("*nooumena*"). So, the world appears, in the way that it appears, as a mental phenomenon. We cannot know the world as a *thing in itself*, that is other than as an appearance within us. If we try to know an object as being other than an appearance, it can only be known as a phenomenal appearance, never otherwise. And a mere phenomenal appearance has not independent existence outside us. This distinction makes it possible an a priori knowledge of perceived objects, that is, a pre-given knowledge that specifies something about them. Furthermore, this division of all objects into *phenomena* and *nooumena* is actually a division of the world within a world of the senses and of the understanding (*mundus sensibilis* & *intelligibilis*).

As far as *scientific knowledge* is concerned, any judgment about the world depends on the cooperation of the senses and understanding.[4] But, reason is "the origin of certain concepts and principles" (Kant 1781/1998, A299/B355) independent from those of sensibility and understanding, and thus, a faculty of principles. All our knowledge begins with the senses, proceeds then to the understanding, and ends with reason. There is nothing above Reason. This theoretical formulation by Kant, especially the distinction between Reason and Understanding, had a great impact on both philosophy and science.

---

[4]Though Kant discusses issues relevant to scientific knowledge and especially Physics in the *Critique*, his views on this topic are developed most explicitly in the *Metaphysical Foundations of Natural Science* (1786).

In summary, our selves shape reality thanks to its equipment, namely, *sensibility*, *intuitions*, *understanding* and *reason*. With sensibility (the senses) we capture from the external world colors, sounds, shapes, odors, etc. These unorganized and 'messy' sense data are then organized by ourselves as representations of the things that are around us, by means of the functions of intuition and reason. Intuition takes also the forms of *time and space*. Our self is placing sense data within the frame of time and space, and thus provide order and relevance. Subsequently, understanding and reason using specific generic concepts or categories, that are inherent to it, turn into objective empirical judgments of things, the individual appearances of our senses which, thanks to intuition, have already incorporated spatiotemporal relevance.

Kant sought to answer the question 'how is science possible' or 'how is the human mind capable of capturing the knowledge of the world?' In doing so, he redefined the relationship between the self/subject and external reality. It is not reality that determines true knowledge. For knowledge to be valid and leading to truth, one's beliefs *must not* be in line with reality but with his/her self. It is the self that finds true knowledge and can call upon it. And this is possible because reality accords with our own self-structure, determining which of our thoughts are true and which are not.

Karl Popper (1902–1994)

Austrian-British philosopher Karl Popper provided an account of knowledge by which he claims that *induction* cannot offer certainty neither achieve it. Starting with Aristotle (Posterior *Analytics*) induction was considered to play a crucial role in the theory of scientific knowledge. With induction scientists start from the complex and heterogeneous empirical data and they manage to master them, discovering behind them general laws and indemonstrable principles. But according to Popper, observing more instances of a general statement may at most, make any general statement more probable.

Popper's argument consisted one of the most influential and controversial views on the problem of induction, and it was presented in *The Logic of Scientific Discovery* (in German 1934/in English 1959).[5] Popper held that induction has no place in the logic of science. Science according to his rather strict view is a *deductive* process in which scientists formulate hypotheses and theories that they test through produced particular observable consequences. Theories are not supposed to be confirmed or verified, but they may be only *falsified* and rejected or *tentatively*

---

[5]The 1934 German title *Logik der Forschung. Zur Erkenntnistheorie der modernen Naturwissenschaft* (Heidelberg: Mohr Siebeck) literally translates as *Logic of Research: On the Epistemology of Modern Natural Science*.

accepted if corroborated and backed-up by all appropriate tests-*in the absence of falsification.*

Following Popper's line of thought, it could be argued that all beliefs about scientific theories are subjective or *personal,* and correct reasoning is merely about how evidence should change one's subjective beliefs over time. Within this framework, some argue (see, for instance, Lipton 2004; McMullin 1992; Psillos 2004) that what scientists actually do is not *inductive reasoning* but rather *abductive reasoning,* or *inference to the best explanation.* In this account, science is not about generalizing specific instances but rather about hypothesizing explanations for what has been observed. Various points of criticism emerged within this approach, mainly concerning: (a) the derivation of truth or its confirmation, (b) the way knowledge comes out of empirical data, (c) the role of a subject in experiencing and how it might be inter-connected among various subjects (inter-subjectivity-based explanation of a theory), (d) the distinction between the objective shape (structure rather than content) of empirical experience and its individual content, (e) whether we should rely on the structural properties or relationships of the empirical experiences in order to form an objective science and (f) the *reduction* of the theoretical terms to the empirical facts (for a critique of *inference to the best explanation,* see van Fraassen 1989).

According to Popper, the central problem in the philosophy of science is that of *demarcation,* i.e., of distinguishing between science and what he called 'non-science'. As already mentioned, he accordingly repudiates induction and rejects the view that it is the characteristic method of scientific investigation and inference, substituting falsifiability in its place. It is easy, he argues, to obtain evidence in favor of virtually any theory. Such 'corroboration', as he terms it, should count scientifically only if it is the positive result of a genuinely 'risky' prediction, which might conceivably have been false (see Thornton 2017).

Popper's *critical rationalism* aims to replace purely justificatory methods with critical ones. As a theory, it is contrasted with the traditional concept of rationality (or in Popper's words *uncritical rationalism*), that is, the view that only what is rationally justified or can be proved by reason and/or experience should be scientifically accepted. Popper argued that uncritical rationalism cannot adequately explain how proof is possible and therefore leads to inconsistencies. "Now it is easy to see that this principle of an uncritical rationalism is inconsistent; for since it cannot, in its turn, be supported by argument or by experience, it implies that it should itself be discarded" (Popper 1977).

Thomas S. Kuhn (1922–1996)

Thomas S. Kuhn's contribution to the Philosophy of Science marked not only a break with several key positivist doctrines, but also inaugurated a new style of *doing Philosophy of Science* bringing it closer to the *History of Science.* His account held that science develops through periods of stable growth punctuated by revisionary revolutions (see Bird 2013).

This historical continuity of science is particularly evident from the very first pages of his best-known book '*The Structure of Scientific Revolutions*', in 1962. "If science is the constellation of facts, theories, and methods collected in current texts, then scientists are the men who, successfully or not, have striven to contribute one or another element to that particular constellation. Scientific development becomes the piecemeal process by which these items have been added, singly and in combination, to the ever-growing stockpile that constitutes scientific technique and knowledge" (Kuhn 1962/1970).

According to Kuhn, a science that enjoys consensus among the scientific community goes through two alternating phases: (a) The phase of *normal science* when the key theories, instruments, values and metaphysical assumptions that comprise the disciplinary matrix are kept fixed, permitting the cumulative generation of puzzle-solutions and (b) the phase of a *scientific revolution*, when the disciplinary matrix undergoes revision, in order to permit the solution of the more serious anomalous puzzles that disturbed the preceding period of normal science (Bird 2013).

Within this disciplinary matrix, consensus on unique instances of scientific research representing examples of good science form, what Kuhn calls, the 'paradigm'. Historical instances that correspond to that description are Aristotle's analysis of motion, Ptolemy's computations of planetary positions, Lavoisier's application of the balance, and Maxwell's mathematization of the electromagnetic field (1962/1970).

Responding to Popper's falsification criterion (critical rationalism) Kuhn holds that during normal science scientists neither test nor seek to confirm the guiding theories of their disciplinary matrix. Although indeed the role attributed to falsification is much like anomalous experiences, i.e., experiences that by evoking crisis, prepare the way for a new theory, the latter are not falsifying those theories. It is rather that these anomalies are ignored or explained away, if possible at all. It is only the accumulation of particularly troublesome anomalies that poses a serious problem for the existing disciplinary matrix (Bird 2013). Consequently, confidence is lost in the ability of the paradigm to solve anomalies a crisis in science. In this context scientists can no longer evade these anomalies that subvert the existing tradition of scientific practice, penetrate existing knowledge to the core and lead to paradigm change (Kuhn 1962/1970).

Finally, Kuhn rejects the idea that knowledge is growing when our theories are succeeding in producing better representations of reality. As Salmon (1992) notes "For Kuhn a scientific theory is better than its predecessors only in the sense that it is a better instrument for formulating and solving puzzles, and not because it is a better representation of what the physical world is really like".

At this point we consider highly valuable and useful to present in brief, the work of two notable philosophers of science who, along with Popper and Kuhn, have greatly contributed to the evolution of the philosophy of science. These are, Imre Lakatos and Paul Feyerabend.

Imre Lakatos (1922–1974)

Following, yet criticizing, Popper, Imre Lakatos attempted to construct a methodology of science whose precepts are more in accordance with actual scientific practice. In doing so, he rejected Popper's demarcation criterion as too restrictive, since it rules out too much of everyday scientific practice as being unscientific and irrational. Scientists often persist—and, it seems, they rationally persist—with theories that by Popper's standards they ought to have rejected as "refuted". One such example is the Newtonian celestial mechanics. And if scientists often persist with "refuted" theories, it is that either the scientists are being unscientific or Popper is wrong about what constitutes good science, and hence about what scientists ought to do. Even though falsifiability continues to play a part in Lakatos' approach, its importance is somewhat diminished. Instead of an individual theory ought to be rejected as soon as it is refuted, the focus of Lakatos is on a sequence of theories constituting what he calls, a *research program* (Musgrave and Pigden 2016).

The nature of these programs and their relationship with Popper's rational criterion are presented by Lakatos in his book *The Methodology of Scientific Research Programmes*. They concern Newton's theory of gravitation, Einstein's relativity theory, quantum mechanics, Marxism, Freudianism, as are being research programs. Each one has a characteristic-stubbornly defended-hard core, a more flexible protective belt and an elaborate-problem-solving machinery together with unsolved problems and anomalies. "All theories, in this sense, are born, refuted and die refuted. But are they equally good? Having described what research programs are like, we may ask how can one distinguish a scientific or progressive program from a pseudoscientific or degenerating one? Contrary to Popper, the difference cannot be that some are still un-refuted, while others are already refuted. When Newton published his *Principia*, it was common knowledge that even the motion of the moon could not be properly explained; in fact, lunar motion refuted Newton. Kaufmann, a distinguished physicist, refuted Einstein's relativity theory in the very year it was published. But all the research programs Lakatos admired have one characteristic in common. They all predict novel facts, facts which had been either undreamt of, or have indeed been contradicted by previous or rival programs... Halley, working in Newton's program, calculated on the basis of observing a brief stretch of a comet's path that it would return in seventy-two years' time; he calculated to the minute when it would be seen again at a well-defined point of the sky. This was incredible. But seventy-two years later, when both Newton and Halley were long dead, Halley's comet returned exactly as Halley predicted" (Lakatos 1978/1989).

As far as scientific growth is concerned, Lakatos emphasizes on a certain continuity that connects scientists' work on every field. This continuity evolves from a genuine research program and corresponding methodological rules. Furthermore, "Even science as a whole can be regarded as a huge research program with Popper's supreme heuristic rule: 'devise conjectures which have more empirical content than their predecessors'...But what I have primarily in mind is not science as a whole, but rather *particular* research programs..." (Lakatos 1978/1989).

Regarding Kuhn's approach, Lakatos expressed particular reservations. He accused Kuhn's position as an appeal to "mob psychology". This is because, according to Lakatos, scientific method conceived as the discipline of rational appraisal of scientific theories vanishes. By Kuhn's way, changes in "paradigms" are explained in terms of social psychology and not governed by rules of reason. As Hacking (1983/1997) notes: "Lakatos utterly opposed what he claimed to be Kuhn's reduction of the philosophy of science to sociology. He thought that it left no place for the sacrosanct scientific values of truth, objectivity, rationality and reason."

Paul Feyerabend (1924–1994)

Paul Feyerabend became known for both for his somewhat unorthodox view of science and his rejection of the existence of universal methodological rules and as a critic of Karl Popper's *critical rationalism*. According to Feyerabend, the idea that science can, and should, be run according to fixed and universal rules (known as *methodological monism*), is both unrealistic and pernicious.

His critique is a *reductio ad absurdum* of a single scientific methodology such as the principle of falsification, the forbidding of ad hoc hypotheses and consistency. In this particular endeavor, he takes the premise that Galileo's advancing of a heliocentric cosmology was an example of scientific progress. Galileo did not follow the *rational* principles of scientific method. Feyerabend argues that, if he had done so then he could not have advanced the Copernican *heliocentrism*. This implies that scientific progress, the new cosmology in this case is actually "a step *back*: apparently relevant evidence is pushed aside, new data are brought in by ad hoc connections" (Feyerabend 1975/1993).

Feyerabend thinks that Galileo could have succeeded in justifying his ideas, not only because he didn't follow strict rules such as verifiability (logical empiricism) or Popper's falsifiability, but also because he broke the rules in some ways. For Galileo, introducing a new idea and a new system against long established Aristotelian physics, meant he would be confronted with much more unverifiable than verifiable facts. Thus, his theory could not be successful at all in competing with Aristotle if it was to be determined by *critical* or *uncritical* rationalism's standards.

So, according to Feyerabend there is no exclusive rule in the procedure of scientific progress. Scientists have to find (invent) ad hoc methods which could fit to their own problems. "To sum up: wherever we look, whatever examples we consider, we see that the principles of critical rationalism (take falsifications seriously; increase content; avoid ad hoc hypotheses; 'be honest'—whatever that means; and so on) and, *a fortiori*, the principles of logical empiricism (be precise; base your theories on measurements; avoid vague and untestable ideas; and so on), though practiced in special areas, give an inadequate account of the past development of science as a whole and are liable to hinder it in the future. They give an inadequate account of science, because science is much more 'sloppy' and 'irrational' than its methodological image" (Feyerabend 1975/1993).

Having been briefly examining the main representatives of Philosophy of Science, we shall now return to the epistemic scheme of Table 1, by presenting the three corresponding components of science as follows.

B1. Theory

Most people understand *theory* as a meaning identifying one's specific belief or concept. But in Science *theory* also refers to the way one interprets phenomena. Nevertheless, the definitions of theory in the Merriam-Webster dictionary provide all following notions:

1. the analysis of a set of facts in their relation to one another,
2. abstract thought and speculation,
3. the general or abstract principles of a body of fact, a science, or an art *music theory*,
4a. a belief, policy, or procedure proposed or followed as the basis of action *her method is based on the theory that all children want to learn*,
4b. an ideal or hypothetical set of facts, principles, or circumstances—often used in the phrase *in theory in theory, we have always advocated freedom for all*,
5. a plausible or scientifically acceptable general principle or body of principles offered to explain phenomena *the wave theory of light*,
6a. a hypothesis assumed for the sake of argument or investigation,
6b. an unproved assumption, and conjecture,
6c. a body of theorems presenting a concise systematic view of a subject *theory of equations*.[6]

All scientific theories set off from a *hypothesis*—a belief assumed to be true, a speculation which is not proven yet. If sufficient evidence backs-up this hypothesis, then one can form a valid methodology to explain this speculation concerning this specific phenomenon, eventually producing a scientific theory.

Natural scientists only agree with very few things about how scientific theories evolve, with an apparent lead to progress. And that is because of a distinct difference between *observation* and *theory*, and the fact that scientific progress is a product of a broader conglomeration of evidence and well-built and rigid procedures. It is commonly believed that terminology should be unique, concrete and clear and that all sciences ought to conform to similar methodologies. It is also

---

[6]Also available at: https://www.merriam-webster.com/dictionary/theory (accessed on 21.4.2017).

agreed that there is a distinct difference between the *context of justification* and the *context of discovery*.

Although simply a hypothesis, it may be argued that a theory may be somehow backed-up by indications, experimental verifications and strict controls, historical conditions and circumstances of the *discovery*. The psychological idiosyncrasies and the economic environment are also significant parameters one must consider in such an account. The contradictions and the refutations of any scientific work gradually in time might shake the established Kuhnian *Paradigm*, as a disciplinary matrix. The eventual gradual challenge and final change of established methods, criteria and basic assumptions should always accompany the certainties invested in any given scientific field. In the words of Nobel-awarded Belgian physical chemist Ilya Prigogine (1917–2003) *"The future is uncertain… but this uncertainty is at the very heart of human creativity"* and the human future is still to be constructed since the Laws of Nature are not given but always remain probabilistic (1996). Riddles will keep challenging existing knowledge and asking to be inevitably solved by the unceasingly intense quest for the scientific truth.

The role of *technological applications* versus *theoretical constructions* is fundamental as much as the accumulation of knowledge and the social role of the scientist. All came into the discussion in the mid-20th century as part of a common project trying to explain the *revolutions* in science, the evolution of new theories and new *Paradigms*, along with their establishment. This account offers any area of science the possibility to an eventual fall and decay before an innovation or renaissance may follow.

Consequently, a theory is scientific if it has *some or even all* of the following features:

(a) It makes testable predictions,
(b) It is falsifiable,
(c) It predicts *new* facts,
(d) It unifies already existing ideas, and
(e) It is consistent with what we already know etc.

However, one should always keep in mind that even a theory that doesn't succeed in meeting one of the aforementioned criteria may be a good or useful theory, since the above is a *non-prescriptive* definition; one only needs be a little cautious about those that fail to meet any or only a few.

B2. Experiment

Justification in physical science is provided through the process of an experiment. Scientific problems are knowledge oriented and may be approached through observations that produce a *data collection* on the phenomena. Understanding nature and thus producing a scientific law describing it, seems to be a simple way of putting it. On the other hand, technological problems are practical approaches; and technology focuses on control of nature by using the *rules* that suggest an *objective way* (?) to exploit nature. This differentiation, defines the further experimentation processes per any given case.

Scientific methodologies have made significant progresses in revealing how *knowledge* of the natural world and phenomena is acquired. Said in scientific terms, a method of inquiry must be based on gathering observable and measurable evidence subject to specific principles of reasoning and experimentation. The scientific method consists of the collection of data through observation and experimentation, and the formulation and testing of hypotheses. "For the qualities of bodies can be known only through experiments; and therefore, qualities that square with experiments universally are to be regarded as universal qualities" (Newton 1687/1999). During the *Scientific Revolution*, Sir Francis Bacon (1561–1626) established and popularized an inductive methodology for scientific inquiry. Within this framework, the use of experiments can correct the much-discussed problem of information gained through the senses. Finally, his famous aphorism, "knowledge is power", is found in the *Meditations Sacrae* (Montagu 1825–1834).

Of course, the way in which a method is considered scientific is not always a simple matter. As Hacking (1983/1997) points out when addressing induction and deduction: "What is scientific method? Is it the experimental method? The question is wrongly posed. Why should there be the method of science? There is not just one way to build a house, or even to grow tomatoes. We should not expect something as motley as the growth of knowledge to be strapped to one methodology." van Fraassen (2008) also implies a high risk of non-objective measuring, but from another point of view, when he notes that: "If we simply put some white powder to the tip of the tongue to check whether it is salt or sugar, we are *making an observation*, conducting a primitive experiment, and in effect *performing a measurement without instruments*. But we cannot take this simple case as very revealing... (in experiments) measurements occur only as special elements of the experimental procedure by which objects are deliberately placed in unusual, artificially designed conditions—conditions in which they are made to respond to the questions put to them. That intricate construction of a well-designed instrumental set-up for experimentation is what we must inspect first, to understand the intricacies of measurement in general".

When making observations, scientists look through instruments equipment and devices from simple ones to more complex, multi-functional and complicated in their operations. Scientists, through the years and having accepted the technological developments and the modern analytical techniques, in general, can firstly agree on what they see and secondly accept what is reported accordingly. Furthermore, all observation involves both perception and cognition. That is, one does not make an observation passively, but rather is actively engaged in distinguishing the phenomenon being observed from surrounding sensory data. Regarding this, van Fraassen (2008) notes: "The heart of an experiment is typically a sort of measurement: the set-up produces or lends itself to a phenomenon that is meant to provide in-formation about the character of some target object, event or process.

The artificially produced or isolated phenomenon is treated as providing data about the target, to provide us with a 'view' of it".

Nevertheless, all of the recorded observations need a further explanation by scientists, who, when ever need to explain a phenomenon, they should adopt and utilize certain theories that on the other hand, may not be the same among them. Therefore, disagreements may develop on what they actually see. Alternatively, whenever there is any suspicion on the object's *picture*, a different theory may come in use. Therefore, observations are affected by one's underlying understanding of the way in which the world functions, and that understanding may influence what is perceived, noticed, or deemed worthy of consideration. In this sense, it can be argued that all observation is theory-laden (see James 2016).

B3. Scientific Truth and Social Construction

As already mentioned, some methods of producing knowledge, such as *trial and error*, or knowing from experience, tend to create a highly *situational* knowledge. Situational knowledge is often embedded in language, culture, or traditions. One of the main attributes of the scientific method is that the theories it produces are *much less situational* and closer to the truth than knowledge gained by other non-systematic methods.

A basic point of situational differentiation of knowledge is the way in which the latter circulates and diffuses within society. The way of creating scientific ideas, research and making knowledge public, refers, in principle, to the process of *circulation* of the ideas and research products in question. The successful circulation of an idea, which is a scientific progression, ends up to the *appropriation* of this progression, of this new theory and application by the societal mechanisms and *institutions*. The appropriation not only affects the cognitive qualities and the cognitive tools of us humans, but also begets to easily traceable institutional changes. A flow of the new knowledge till its embedment by the social, financial and political institutions may be depicted as follows.

Ethical and value-issues arise during the transfer of new knowledge to the scientific journals, along with its presentation to congresses, various fora, and certainly the Web and the media. This transfer of knowledge comprises a process known as *social constructivism* before its final appropriation by society. Issues of meaning and cultural changes arise during the second phase of the transfer; these are not merely a new knowledge, but a socially constructed and linguistically fermented 'cognitive product'. A philosopher of science should also concern to address suggestions to journalists and lab directors on how to enhance their performance in terms of popularization and knowledge transfer to avoid scientific distortions and inaccuracies due to private or third-party interests in general, which is a rather significant ethical and epistemological issue (Theologou 2014).

Regarding truth, three distinct approaches have been developed within the scientific community, namely:

(a) The *Realists* who claim that science aims at *truth* and that one should regard scientific theories as true, approximately true, or likely true. Realists often point to the *success* of recent scientific theories as evidence for the truth (or near truth) of current theories. Scientific realism argues that entities, conditions and procedures that are described by correct theories, do exist. Even if existing scientific capabilities have not yet managed in full, it is believed that they are approaching the truth, aiming at revealing the mostly inner-structure of things and phenomena.

(b) The *Anti-Realists* who argue that science does not aim (or at least does not succeed) at *truth*, especially truth about non-observables such as electrons or other universes. Antirealists point to either false theories in the history of science, success of false modeling assumptions, or widely termed postmodern criticisms of objectivity as evidence against scientific realism. They attempt to explain the success of scientific theories without reference to truth. Some antirealists claim that scientific theories aim at being accurate only about observable objects and their success is primarily judged by that criterion. Anti-realism holds that there are no theoretical entities, although the relevant phenomena do exist. The reason for *constructing theories* is merely for predicting and creating facts within the area of our interests; i.e., theories on non-existing entities enhance our ways of thinking, and despite their adequacy and need for using them, they are not true. Models are still useful in working and arranging things in our mind, but they do not depict *true pictures* or *representations* of nature. One may feed in the appropriate data and acquire the *right* outcome, but even by doing so still nothing affirms that one deals with the proper natural phenomenon in hands.

(c) *The Instrumentalists* who argue that scientific theories should only be evaluated on whether they are *useful*. They claim that whether theories are true or not, is beside the point, because the purpose of science is to make *predictions* and enable effective technology.

There are further considerations than the difficulties of defining the meaning of *true* or *real*. They concern the role of philosophy in ratifying this and not another account, the role of laboratory equipment or other measuring apparatus, the reasons and their contribution in understanding an *entity*, the concept of *knowledge* itself and the role of *experience*, along with particular questions deriving from certain scientific fields. Overall, the question about theories may be summarized in whether they are *real* or *false*, whether they have a strong proposal to the truth or whether they are targeting to truth.

As far as entities is concerned, the only question is whether they exist. Usually, if a theory is accepted as true, so are the relevant entities, as the opposite would make no sense. Still, someone may be realist regarding the entities, but antirealist regarding a theory, something quite common in religions and the acceptance of god. Realism about entities argues that a good many theoretical entities really do exist. Anti-realism denies that by claiming that they are fictions, logical constructions, or parts of an intellectual instrument for reasoning about the world. Or, in a less dogmatic way, we have not and cannot have any reason to presume they are not fictions. We do not need not assume their existence in order to understand the world. "*Realism about theories* holds that scientific theories are either true or false regardless of what we know: science at least aims at the truth, and the truth is about how the world is. Anti-realism claims that theories are at best warranted, adequate, good to work on..." (Hacking 1983/1997).

C. Modeling and Scientific Representation

Early in the 1980s, the Canadian analytic philosopher Ian Hacking (1936–) advocated rethinking the role of hands and eyes in constructing scientific knowledge. His *Representing and Intervening* is one of the most quoted philosophical books that attracted an urgent paying attention to experimentation, because the latter "has a life of its own" (Hacking 1983). He was influenced by debates on the standpoints of Thomas Kuhn, Imre Lakatos, Paul Feyerabend and others, and he is notably known for bringing a *historical approach* to the Philosophy of Science.

Scientific theories do not consist of a mere conglomeration of evidence. They are models of the world that in a certain extent may be tested by evidence. Testing does not define theory in an absolute manner, since humans intervene through their working environment and other societal values making thus any theory socially and culturally defined. This doesn't point to an arbitrary or autonomous course of action, but brings the discussion to the way scientists take advantage of instruments, measurements and the findings of their experiments in order to convince the rest of their colleagues about the *superiority* of their own programs. Since there are various ways of both promoting and developing the research and inventing a corresponding functional model, these two fields cannot but to be depended on the perspective of each scientific group. Their success also contains the capacity of their group to convince the others. Considering the miscellaneous standpoints and aspects, one may see into the difficulties in objectively defining science and framing its results and remarks about a phenomenon, system or concept.

Let's now follow the context of the idea that all observation is theory-laden. It is widely acceptable that Mathematics, especially, *applied mathematics*, has offered a universal language describing and/or modeling physical problems in a unique, adequate, sufficient and concrete way. Mathematics can progress a theory even further by offering *predictions* instead of *descriptions*. This derives from the intrinsic ability of mathematics to obtain solutions in a space of parameters and conditions, otherwise unattainable either in real world observations or in laboratory scale experiments. Beyond that, applied mathematics originated tools that have the potential to be independent from the application for which they have been developed. Furthermore, these tools can *describe, explain and predict* the phenomena and situations far beyond the initial application.

The mathematical description of a system (as well as its solution) under initially specified and boundary conditions, requires a pre-existing knowledge of not only of the physical, chemical, and other processes under description, but also of the suitable mathematical theories, expressions, norms and techniques. Mathematics is also being developed in time widening the potential for descriptions and prediction of natural phenomena, therefore its significant contribution to understanding the world around us. However, it is also clear now that this knowledge strongly affects the accuracy of the description, the precision of any solution and the physical interpretation of the results. In this context, mathematics seems, up to a certain extent, like a quasi-distorting lens through which we see reality rather than an objective, hence, an independent tool of the *new knowledge* creation.

Representation

Science provides us with *representations*. And it is through these representations that humans learn about the world. In this context, philosophers of science have long acknowledged the importance, if not the primacy, of scientific models as representational units of science (Frigg and Nguyen 2016). Nowadays, mainstream model theory is a sophisticated branch of mathematics, but in a broader sense, it is the study of the interpretation of any language, formal or natural, by means of set-theoretic structures. It concerns the representation of how language describes the actual world or any possible world. Sir Francis Bacon had already identified the problem of language. Language and words often betray their own purpose leading to an obscuring of the very thoughts they are designed to express. They often carry false significance, leading to errors, logical fallacies or in Bacon's account *Idols of the Marketplace*. "But the idols of the marketplace are the biggest nuisance of all, because they have stolen into the understanding from the covenant on words and names. For men believe that their reason controls words. But it is also true that words retort and turn their force back upon the understanding; and this has rendered philosophy and the sciences sophistic and unproductive" (Bacon 1620/2000).

The symbolic language developed by mathematics and engineering is meant to give meaning and clarify ambiguities. Associated with any formal language, each model consists of a set (the domain of the model) and special subsets of the domain. The domain is a mathematical representation of the objects in some possible world, and the subsets of the domain are "mathematical representations of properties or relations among things in that possible world. The idea, roughly, is that a property corresponds to the set of things that have that property in each possible world. 'Brown' corresponds in each possible world to the set of brown things in that world" (Salmon 1992). Likewise, Salmon (1992) provides an example of scientific representation, indicating the notion of *coordination*: "Let us select as the model for our theory the manifold of 'all real numbers R. Each real number in R represents a particular instant. This representation relation is a *coordination* of the instants of the physically possible world with the mathematical structure R so that the relation is commonly called a *coordinate system*. We can infer many of the temporal properties of the physically possible world from the coordination".

More specifically, all scientific theories are typically presented with mathematical equations that involve specific terms. One classic example is the *ideal gas law*: $PV = nRT$, where $P = pressure\ of\ the\ gas$, $V = volume\ of\ the\ gas$, $n = amount\ of$

substance of gas, R = universal gas constant and T = absolute temperature of the gas.
Although these terms are functional symbols with specified mathematical character, they
are often pronounced as if they were entities already familiar before the theory's
introduction. Other examples are 'temperature', 'volume' 'distance', 'time', 'pressure',
and so forth. van Fraassen notes (2008) that any "theory would remain a piece of pure
mathematics, and not an empirical theory at all, if its terms were not linked to measurement procedures. But what is this linkage?" The answer to this question concerned
philosophers of science of early 20th century and highlighted the so-called *problem of
coordination*. Beginning with Ernst Mach's (1838–1916) discussion of the relation
between mathematical and physical geometry, the problem of understanding just how a
scientific theory is more than its mathematical guise came to special prominence through
the writings of Moritz Schlick (1882–1936) and Hans Reichenbach (1891–1953).

In 1920, Reichenbach writes: "It is characteristic of modern *physics* to represent all
processes in terms of *mathematical* equations. But the close connection between the
two sciences must not blur their essential difference.... The *physical object* cannot be
determined by axioms and definitions. It is a thing of the real world, not an object of the
logical world of mathematics. Offhand it looks as if the method of representing physical
events by mathematical equations is the same as that of mathematics. Physics has
developed the method of defining one magnitude in terms of others by relating them to
more and more general magnitudes and by ultimately arriving at "axioms", that is, the
fundamental equations of physics. Yet what is obtained in this fashion is just a system
of mathematical relations. What is lacking in such system is a statement regarding the
significance of physics, the assertion that the system of equations is *true for reality*
(Reichenbach 1965 *in* van Fraassen 2008).

Focusing on a more general notion of the concept, each scientific theory has two
forms of artificial representation: (a) concrete (graphs, scale models, digital displays
etc.) and (b) abstract (mathematical models). The specific features of the artifacts
used determine the entire theoretical construction. The problem of coordination or
how the represented must be like its representation inevitably refer to these elements, but also to the way any science relates to its domain of application.

From the above it becomes evident an underlying need to understand and accept
a functional model that could also promote scientific evolution. A model based on
the premise that science really offers highly elaborate knowledge of how the world
works and takes also into account the bias and the variability that a *true* modeling of
nature implies. The evolving of science could arise from an understanding of what
exactly prevents research from giving immediate answers to any complex problem,
without diminishing the prestige and integrity of the scientific method. In our view,
this need manifests itself particularly in the field of *Engineering*.

D. The Case of Engineering

Engineering aims at modifying the natural environment through the design and
manufacture of artifacts. In this sense, it could be contrasted with science, the aim
of latter be to understand nature. Furthermore, what distinguishes engineering

design from artistic design is the requirement of the engineer to make more accurate quantitative predictions of the behavior and effect of the artifact prior to its manufacture. Due to the great impact of the exercise of this profession on society, engineering could be considered a social and not merely technological discipline.

Since what engineers do is subject to individual or collective evaluation, philosophical issues as far as its application is concerned, emerged. Engineers are regarded as being motivated by the challenge of being able to do something that has not been done before. Their actions include extending human control over Nature and developing powerful and efficient devices, whether there is a practical need for these or not. "Engineering can be seen as an expression of an inherent human need to overcome uncertainty and fear of the unknown, and to be able to dominate and exploit our environment. It may be the same force that drives evolution..." Aslaksen (2017). Assumptions like this led to the development of the *philosophy of engineering*. This area deals with issues such as objectivity of experiments, engineering ethics in the workplace and in society, the aesthetics of engineered artifacts, etc. In general, philosophy of engineering is an emerging discipline that considers what engineering is, what engineers do, and how their work impacts on society.

The book you have in your hands refers firstly to experimentation in engineering; in other words, to the theories, means and techniques being involved in the investigation of particular phenomena. For engineering as well as for other sciences, experimenting is actually a reconstruction of the phenomena in a controlled environment. A reconstruction for testing a particular hypothesis via a justified intervention to the selected conditions that could allow the phenomena to evolve. It also focuses on what experiment is and what experimenters do from an engineering point of view. A special attention will be given on what is actually represented when an experiment is designed and re-constructed based on existing knowledge. Therefore, this introductive chapter has been organized in rather distinct but highly relevant sections of contents. Beginning with the central notion of knowledge, to science as a whole and to particular areas pictured through the years by philosophers of science, and concluding to a portrayal of the phenomena and their perception/representation. What we wanted to highlight was a general illustration of how engineering is part of the epistemic scheme that supports science in its entirety. A close relationship between engineering and existing or produced knowledge about phenomena is obvious and this is attempted to be represented by a mathematical model.

References

Aslaksen EW. Engineers and the evolution of society. In: Michelfelder DP, Newberry B, Zhu Q, editors. Philosophy and engineering exploring boundaries, expanding connections. Switzerland: Springer; 2017 pp. 113–24.

Bacon F. The new organon. In: Jardine L, Silverthorne M, editors. Cambridge: Cambridge University Press; 1620/2000.

Bird A. Thomas Kuhn. In: Zalta EN, editor. The Stanford encyclopedia of philosophy. 2013. https://plato.stanford.edu/archives/fall2013/entries/thomas-kuhn/.

Feyerabend P. Against method. London: Verso; 1975/1993.

Frigg R, Nguyen J. Scientific representation. In: Zalta EN, editor. The Stanford encyclopedia of philosophy. 2016. https://plato.stanford.edu/archives/win2016/entries/scientific-representation/.

Gärdenfors P. Belief revision: an introduction. In: Gärdenfors P, editor. Belief revision. Cambridge: Cambridge University Press; 1992. pp. 1–28.

Gettier EL. Is justified true belief knowledge?. Analysis. 1963;23:121–123. Full paper also available at: http://www.ditext.com/gettier/gettier.html. Accessed 21 Apr 2017.

Hacking I. Representing and intervening. New York: Cambridge University Press; 1983/1997.

Ichikawa JJ, Steup M. The Analysis of Knowledge. In: Zalta EN, editor. The Stanford encyclopedia of philosophy. 2017. https://plato.stanford.edu/archives/spr2017/entries/knowledge-analysis/.

James DD. Neo-classical physics or quantum mechanics? A new theory of physics. Educreation Publishing; 2016.

Kant I. The critique of pure reason. Translated by Guyer P, Wood AW. Cambridge: Cambridge University Press; 1781/1998.

Kant I. Metaphysical foundations of natural science. In: Friedman M, editor. Cambridge: Cambridge University Press; 1786/2004.

Koutoungos A. About philosophical method. Athens: Pedio; 2012 (in Greek).

Kuhn TS. The structure of scientific revolutions, 2nd ed. Chicago: University of Chicago Press; 1962/1970.

Lakatos I. The methodology of scientific research programmes. Philosophical Papers, vol. 1. New York: Cambridge University Press; 1978/1989.

Lipton P. Inference to the best explanation, 2nd ed. London: Routledge; 2004.

McMullin E. The inference that makes science. Milwaukee, WI: Marquette University Press; 1992.

Montagu B. The works of Francis Bacon, Lord Chancellor of England. A new edition, vol. 1. London: William Pickering; 1825–1834.

Musgrave A, Pigden C. Imre Lakatos. In: Zalta EN, editor. The Stanford encyclopedia of philosophy. 2016. https://plato.stanford.edu/archives/win2016/entries/lakatos/.

Newton I. The Principia: mathematical principles of natural philosophy, Translated by Cohen IB, Whitman A. Berkeley: University of California Press; 1687/1999.

Nozick R. Philosophical explanations. Cambridge, MA: Harvard University Press; 1981.

Popper K. The logic of scientific discovery. London: Hutchinson & Co.; 1959.

Popper K. The open society and its enemies, vol. 2. London: Routledge; 1977.

Prigogine I. La fin des Certitudes. Temps, chaos et les lois de la nature. Paris: Odile Jacob; 1996.

Psillos S. Inference to the best explanation and bayesianism. In: Stadler F, editor. Induction and deduction in the sciences. Dordrecht: Kluwer; 2004. pp. 83–91.

Quine WVO. Two dogmas of empiricism. Philos Rev. 1951;60(1):20–43.

Reichenbach H. The theory of relativity and a priori knowledge. Translated by Reichenbach M. Berkeley: University of California Press; 1965.

Russell B. The problems of philosophy. Oxford: Oxford University Press; 1912/2001.

Salmon MH, et al. Introduction to the philosophy of science. Indianapolis/Cambridge: Hackett Publishing Company; 1992.

Schwitzgebel E. Belief. In: Zalta EN, editor. The Stanford encyclopedia of philosophy. 2015. https://plato.stanford.edu/archives/sum2015/entries/belief/.

Stang NF. Kant's transcendental idealism. In: Zalta EN, editor. The Stanford encyclopedia of philosophy. 2016. https://plato.stanford.edu/archives/spr2016/entries/kant-transcendental-idealism/.

Stelios S. Approaching the other: investigation of a descriptive belief revision model. Cogent Psychol. 2016;3:1164931.

Theologou K. Publication of the scientific progress: some aspects of ethics and culture. In: International Conference: In and Out of the Lab. Science and Technology in the Public Sphere,

13–15 March 2014. Abstract available at: http://inoutlab.web.auth.gr/eng/wp-content/uploads/2013/12/InOutLab-Abstracts-GRENG.pdf Accessed 25 Apr 2014.

Thornton S. Karl Popper. In: Zalta EN, editor. The Stanford encyclopedia of philosophy. 2017. https://plato.stanford.edu/archives/sum2017/entries/popper/.

van Fraassen B. Laws and symmetry. Oxford: Oxford University Press; 1989.

van Fraassen BC. Scientific representation: paradoxes of perspective. Oxford: Oxford University Press; 2008.

Scientific Research—Perspective, Awareness and Criticism

Antonios Kanavouras, Frank A. Coutelieris, Kostas Theologou and Spyridon Stelios

What is the initiative and the virtual impulse for seeking *new* knowledge? Is there a significant differentiation between basic and applied research when it comes to learning the truth? How much does the society trusts the scientific truth and what is the impact of science on the social, technological and ethical development of a society? Do all the above imply that science is providing certainty? Yet, certainty is no more an issue, since the end of certainties was already *established* by Prigogine 1996).

This chapter aims to investigate the way science has been working around the phenomenon, how these have been analyzed and which are the potential criteria, in a narrow and a broader sense, that have been used for assessing its substance and expose the results to the rest of the scientific community for approval and acceptance.

Introduction

The way scientists mainly perceive a phenomenon is in a direct dialectic relationship to the context of the upcoming hypothesis. That needs to have a straight-line reference to a specific question made about this phenomenon per se or on a futuristic version of this, a vision of how things may differ and if and how could that be possible and functional. At this point a note on the differences between scientific and technological research could be made to correlate the outcome between possibility and functionality, respectively. Furthermore, it is implied that the hypothesis may be formed either as a "descriptive" goal or a "predictive" one. Nevertheless, the initial "definitive portrayal" of the phenomenon that establishes the base line for either of the aforementioned goals, has an apparent dependency of what each one of us anticipates regarding this phenomenon, i.e. what we hypothesis that this sensed natural development will do (for us).

© The Author(s) 2018
F. A. Coutelieris and A. Kanavouras, *Experimentation Methodology for Engineers*, SpringerBriefs in Continuum Mechanics, https://doi.org/10.1007/978-3-319-72191-0_3

Hence, a hypothesis is, to a large extent, a human construction supported by the personal interpretations, cognitional formulations, understanding and questioning, which comes together in the form of an "interest". In that sense, the hypothesis is created within a certain "atmosphere", through which the expressions utilized by humans for presenting their interest and the relevant ways for a satisfactory road map to an outcome, are perceived. The goal of the hypothesis to find its potential expression goes via the logical combination of empirical observations and conclusive theoretical and practical interpretations. Thus, incidents occurring in the real world could be translated to phenomena when recognized through the human senses (Kanavouras and Coutelieris 2017c) and placed within the framework of theory and knowledge, available at the very specific time (Kuhn 1962). Consequently, at the main drivers for scientific and technological developments, may to a great extend be defined as the hypothesis formation, the particular interest of the human beings, the related means to satisfy their interest and the outcome of the research applied for that purpose.

Before discussing on the *perception of a phenomenon*, it is crucial to acknowledge and take into account Prigogine's claims about time, chaos and the laws of nature. According to Newton, nature's laws provide a descriptive account of a time-less, deterministic universe, where predictions are dictated by absolute certainty. If time is a fallacy and/or time is reversible, then it has no meaning to deal with any predictions. In this context, Ilya Prigogine's pioneering research led to radical changes. Scientists always discover inconsistencies, fluctuations and instabilities leading to evolutionary formations, in all fields, from cosmology to moral biology. Time reversible processes are rare in the real world; non-reversibility seems to be the rule. No one prior to Prigogine had ever managed to take into account this rather obvious, non-reversible time flow in Physics' laws. In *La fin des certitudes* Prigogine combines concepts he had previously introduced in his works in an austere, coherent scientific worldview.

A phenomenon cannot be independent in its perception from a theoretical scheme that engulfs the phenomenon within a structurally holistic scheme. At the same time, the theories are supporting and are being supported by the phenomena. At a consequent stage, a current theory may hence, explain why some phenomena occur (or not) by modeling the parameters, causes and conditions that frame their occurrence (or non-occurrence) for the reasons of any experimental prediction and control. Alternatively, a theory may explain a *lawful* regularity among empirical events by producing a theoretical/cognitional model of the causes or conditions that, if fulfilled, necessitate the lawful regularity among these events.

Theoretical issues placed by a researcher, which are being formulated into sharp and accurate "technological" questions, may require reproducing the phenomena within the lab, in order to mirror theory versus reality according to the re-produced corresponding expressions of the phenomena under given conditions. The point is that an experiment created within a given lab "environment" has a unique outcome, which should be understood—to a certain extent—against an existing or an emerging, theoretical scheme. Within this process, it is the experimenter, who should produce specific "technological devices"/experiments, through which a decisive answer to these questions, (hypothesis), has to be elicited. More pending issues, rifles and open points, may also follow a gradual implementation into the way the experimentation will be unfolding within its space and time frame. An obvious impact on the research subjectivity and its outcome may now be acknowledged.

Through a following up step of an explanation and interpretation process, the human engagement is essential for understanding what has actually happened during the experiment. Humans, work towards that by providing the expressions of governing principles, while, at the same time, humans also try to make any more sense of themselves as "professionals", their world and the mode of being part *in* it (Popper and Eccles 1977).

By summing up the individual characteristics being examined and "adding" them up into a total picture for obtaining a comprehensive research process, may not seem appropriate when the total systems' behavior is to be understood as dialogic, and emerging through its inherent interaction(s) between self and other parties (Goffman 1959).

The aforementioned discussion concerns the struggle towards the creation of additional valid and justified knowledge parts. Although the form of the creation, presentation and criticism of these outcomes has been developed to its side forms through the years, the core of this justification process remains mainly stable and quite standardized within the various scientific fields. On the other hand, the rhythm of knowledge production has been increasing in size and quality in parallel to the technological research-assisting, developments, leading to enormous data accumulated reports, papers and information, asking for either equally developed management systems, or increased selectivity and quality versus quantity. Nevertheless, the way humans have been selectively collecting the existing knowledge parts available for a given system includes the objectivized pre-understanding, as well as an interpreter and an inquirer. Potential users of the available scientific knowledge parts may be sharing a theoretical and practical pre-understanding within the professional communities. This consequently deriving variability may be defining the multiple horizons of pre-understanding, while understanding also occurs as an iterated reciprocal movement between (the meaning of) a part and (the meaning of) a whole to which that part belongs. Assuming that a part only makes sense within a whole, yet the whole does not make sense except in terms of a coherent configuration of its parts (Gadamer 1994). Both the individual and the whole do not remain independent of each other or unaffected by any relevant conceptualization development which dialectically combines both through the evolution process. A historical reference to any scientific area may easily support this point since within any scientific paradigm a coherent relationship has been developed that had either strengthen this paradigm or had led to its abolishment and the consequent change of the paradigm. In that sense, a dialectic relationship is established beyond the existing of up-to-day concepts, but in a broader extend, with those to come as well, since they question each other a priori. The mode and nature of these dialectic relationships should a certain extent defined the evolution of the history of science and consequently the history of human civilization. The dialect of any invention to the societal background of its time is an interesting example in asking "what if". What if the Greeks had invented the steam power utilization 2000 years ago and had applied that into a steam turning wheel? How would the industrial revolution, if at all, had been developed since then? Where would the human civilization have been by now? Why this did not happen? Initially seemingly a pure metaphysical question, it has nowadays become of interest to historians of science and technology to respond to such challenges, in an attempt to recognize the drivers behind evolution and investigate the parameters and conditions that had altered and keep on altering, the "progress".

Finally, "understanding" contains the information-derived-knowledge. Understanding contains specific knowledge about a subject, situation, etc., or about how something works. Besides the ability to appreciate something, understanding also refers to the human power to abstract thought. It's the individual's perception or judgment of a situation.

Therefore, knowledge has an inherent "engineering functionality" mechanism which transforms the existing knowledge, through means of appropriate justification, into (scientific) *understanding*. That, according to Capurro (1987), belongs together with technology, to the will-to-power as knowledge, to the will-to-know or, in other words, to a "technology of/for knowledge". Furthermore, the hermeneutical paradigm offers a framework for the foundation of various relevance criteria such as systems relevance and individual relevance or suitable applicability (Lancaster 1979; Salton and McGill 1983). Nevertheless, this distinction is not enough. According to Froehlich (1994), hermeneutics can provide a more productive framework for modeling systems and user criteria. This framework should include a hermeneutic of (a) users, (b) information collection, and (c) mediation through the system. Yet, even when this happens, the process of interpretation is essential for the constitution of meaning (Capurro 1985).

In order to overcome any issues raised by the complexity of the phenomena, the human factor engagement and the data collection, in this book we propose an independent, knowledge-engineering based method, allowing the experimenter scientist to design and perform the reproduction of the phenomena in the lab, via a clearly defined experimentation "device". The background of this methodology will be further unfolded within this as well as within the consequent chapters of this book.

A. The World of Phenomena

Etymologically speaking, *phenomenon* has a long history deriving from ancient Greek and the verb *"phenome"* (φαίνομαι, I appear); it was also opted for expressing philosophical opinions on visual incidents and reality. An important background in using the term implies that a phenomenon has to be observable, visible and particular, that appears frequently under given conditions. It also signifies a unique and rather important event. Its regularity allows us to express the phenomenon with a normative *generalization* that approximates a law, turning it into a *universal truth*, thus scientific knowledge. In many instances, this specific regularity is called phenomenon. Still, phenomena need to be *resolved*. But could that be possible with descriptive laws and can we reveal the causes behind the phenomena?

Being in such a close encounter with the human senses, a few ancient philosophers thought of the phenomena as transformed objects of senses, in contrast to the pure substance of things, the permanent reality. In the later years, phenomena were opposed to cognition products, "things as are". Philosopher Kant (1724–1804) moved that discussion to the *modern science* of his times and declared the 'things as they are in themselves' (*noumena*) as not-known and the physical science was defined as the science of the *effects* rather than that of the phenomena, following the normative framework of positivism or the conceptual toolkit of phenomenology (Kant 1781/1998).

Nevertheless, reported phenomena and their corresponding recorded effects have been progressively accumulated in science, being mostly accepted solely on being following experimentally affirmative indications. Based on that, we may distinguish the phenomena as the events which can be recorded by a trained, therefore skillful and competent, observer, who is, then, not interfering in the phenomena, while their recorded effects connect the phenomena to the experimentation process. Through which later process, the phenomena become known and may be described. The way these effects have been recorded and reported through research and outcomes heavily involving humans, eventually makes them the products of a significant human interference to nature. It is apparently part of the scientific research application purpose to attempt a normative norm or *regularity*. At a first glance, this may be seen as acceptable or non-acceptable by the scientific community only when placed against a given theoretical background.

Consequently, creating an event at the lab, denotes an invention skill, since it involves tools, fabricated devices, means of control, the surrounding space or environment, articulated research and successful applications, in some cases within more than a merely one innovative series of reproductive processes. As mentioned above, the experimentation as much as the hypothesis are being developed, evolved and potentially be concluded, within a certain "atmosphere" characteristic of the particular lab. The "lab" then becomes more than a battle field for research; it is a complete micro-world mapped for expertise, equipment and facilities. But, does this then might imply that particular phenomena and events may only form *a reality* within a certain device and human, as well as, technological background? Does it mean that a researcher in a lab creates his/her own reality, along with the experts around, that may not even be reproducible or repeatable, thus, particular and incidental in time and space? How may such a research be validated and accepted by others?

When scientists are forming and collecting experiences, their perceptions are coming together only contingently, so that no necessity of their connection exists or can become evident within the perceptions themselves. Apprehension is only an association of the manifold of empirical intuition, therefore no representation of the necessity for a combined existence of the appearances that it compares, in space and time, is to be encountered in it. Then, apparently experience encounters the cognition of the objects through our senses and hence human perception. So, the relation in the existence of the manifold is to be represented in it not as it is juxtaposed in time but as it is objectively conceptualized in time. Yet, since time itself cannot be perceived, the determination of the existence of objects in time can only come about through their combination in time in general, therefore, only through "a priori" connecting of concepts. Now since the objects carry along their necessity for existence, experience may thus be possible only through a representation of the necessary connection of the perceptions of the objects.

Research as a Concept

Through the centuries, some of the most prominent philosophers have consented that if something is error-free, then it is logic! But where logic derives from? Could we ever reach the *bottom* of logic? Is there an answer to every question? In 1931 an unpredictable response was given; Austrian mathematician Gödel (April 28, 1906–January 14, 1978) at his 25 proved the inadequacy of logic. Mathematics in a peculiar way begins earlier than logic, while the logic's tinder isn't originated from a divine truth but from very fundamental human acknowledgements!

One of the most precious chapters of Philosophy is Gödel's famous *incompleteness theorems* (Gödel 1931).

1. *If the system is consistent, it cannot be complete.*
2. *The consistency of the axioms cannot be proved within the system.*

What does consistent and complete might mean? A logical system is *consistent* when it lacks *contradictions*, when a sentence cannot be *true* and *false* simultaneously. Complete is a system when all its sentences are either false or true.

Therefore, Gödel proved that a system satisfying both theorems cannot exist. If it is complete, then it cannot be consistent and vice versa. In simple terms, it will contain either *contradictions* or questions that cannot be answered.

In his second theorem, the Austrian philosopher clarified that axioms cannot sufficiently prove the consistency of a theory. Hence, Logic that mathematicians through the centuries had been striving to identify with mathematics accepted a horrible and incurable wound. It soon proved as the Achilles' heel of mathematics, turning it quasi incompetent to describe science in whole. The birth of mathematics and of any other science doesn't kick off from absolute logic but from *intuition* itself!

The Human Factor

Even more nowadays, a human engaged in the scientific research, is not an isolated inquirer trying to reach others or the outside world from his or her encapsulated mind/brain, but is someone already sharing the world with others. Modern knowledge is supposed to be, shared by the scientific community. Hence, modern knowledge may emerge not as the primacy of rational or scientific thought that is in qualitative terms superior to all other types of discourse, neither via human subjectivity, as opposed to objectivity, in which inter-subjectivity and conceptuality play only minor roles, not even as the Platonic idea of human knowledge being something separate from the knower.

Acquiring knowledge in the modern way, via a system, includes on the one side the objectivized pre-understanding and on the other side the interpreter or inquirer. Potential users of scientific knowledge are sharing a theoretical and practical pre-understanding with, for instance, professional communities; that is defining the various horizons of pre-understanding. Among those, in a later chapter of this book, we shall further discuss the similarity of knowledge through an appropriated classification scheme.

Scientific research seeks explanations by systematically using a predefined set of procedures in order to collect evidence and other data not determined in advance, along with findings that are applicable beyond the limited framework of the study. In that concept, qualitative research shares these characteristics, especially effectively for obtaining culturally specific information about the values, opinions, behaviors, and social contexts of particular populations. It seeks to understand a given research problem from the perspectives of the local population it involves (Mack et al. 2005). The strength of qualitative research refers to its ability to provide complex textual descriptions of how people experience a given research issue. It provides information about the "human" side of an issue—that is often the contradictory behaviors, beliefs, opinions, emotions, and relationships of individuals. In a similar vein, qualitative methods are also effective in identifying intangible factors, such as social norms, socioeconomic status, gender roles, ethnicity, and religion, whose role in the research issue may not be readily apparent. Finally, when used along with quantitative methods, qualitative research can help us to interpret and better understand the complex reality of a given situation and the implications of quantitative data.

Human engagement of the world of phenomena occurs in multiple domains (Roth 1987). As material entities or living organisms in a physical world, we give expression to basic principles such as that of motion. Yet another domain of human engagement is the domain of the sign, a *semiotics* domain where humans try to make sense, to conceptualize within a structured and realistic model themselves, their world, and the manner of being in it by means of signs and symbols (Peirce 1931–1936, 1958; Popper and Eccles 1977). When engaged in scientific processes, humans carry along the aforementioned occurrence. Yet, as researchers and analytical thinkers, they want to know not only the characteristics of the "real" (the rules of the nature in the case of physical science) but also the practices by which the real is produced and sustained.

Through the historic process of humans' understanding, a focus on "relationships" rather than "separate entities" was developed. Within this context, summing up the individual characteristics being examined and "added up" to make the whole, has not been considered as an appropriate skill, compared to total systems' dialogic behavior emerging in the interaction between self and other participants (Goffman 1959; Miller 1982).

According to Anderson (2009) the real world can be divided into two parts, namely the one constructed of material conditions and the other one of social practices. In that sense, laws would have immediately change if or when, the scientific community had decided on what it considered as a better *explanation*. A scientific field can be formed through the twin principles of public presentation and critical review, or "critical rationalism", expressed by Popper (1959/2002) and what might now be called *public* (rather than self) *reflexivity* (Kobayashi 2003). A publicly reflexive science (one that looks back on itself as an object of study), accepts its constituting action and enters the objectivity struggle to attain the universal knowledge.

Although qualitative findings are often extended to phenomena with characteristics similar to those in the study, gaining a rich and complex understanding of a specific empirical context or phenomenon, typically takes precedence over eliciting data that can be generalized to other scientific areas or applications. In this sense, qualitative research differs slightly from scientific research in general, but it is still a type of scientific research. As Anderson notes (2009), qualitative research "is used to indicate a set of text-based or observational methods that are themselves used as companions to quantitative methods (Wilk 2001). It is used to point to an independent set of methodologies that can be used with or without quantitative methods, but remain within the same epistemological framework (e.g., Chick 2000). Finally, it is used to designate an entirely different paradigm of science that is not only independent of quantitative methods but also of its epistemological foundation (Denzin and Lincoln 1994).

In *metric empiricism*, a term coined by Anderson (2009) for quantitative research, explanation is located in the individual (an epistemological requirement known as methodological individualism). This means that everything needed to explain some activity of the "individual" must be found within the boundaries of the individual. Metric empiricism works from a metric logic of quantities and rates, similarities and differences, dependence and independence, operations and results. It depends on "things" which have clear boundaries (such as "thing" and "not thing") and mutually exclusive characteristics (such as "thing 1 not thing 2"). One can count things, measure their boundaries, figure their proportion among other things, and so on to create all sorts of information found in the metric empiricism explanation. This requirement for achieving good explanations is why such theoretical structures as attitudes, scripts, and schemata have developed: because they are contained within the mind of an individual and not yet through direct scientific practices. In accordance with typical cognitive theory, that constitutes the engine for most of the metric empiricism, while differences have to do with the characteristic explanation that is produced by each form (Anderson 2009).

In the same context, Anderson (2009) also uses the term "hermeneutic empiricism" (HE) to underscore the paradigmatic nature of quantitative research. The hermeneutic empiricist not only tells the story but also interprets it. In the process of employing the HE, she then works from a narrative logic of routines and actions, critical instances and episodes, conversation and discourse, results and phenomena. For the narrative logic to deliver, the quality of the interpretation needs to be proven. Thus, interpretations are based on agents that both generate the action and depict some cultural understanding. They have recognizable action of a beginning, a middle and an end, and last but not least they have motive, intentionality, and consequence. With the goal of arriving at a final statement, a transcendental law needs to hold for all conditions, within the scope of the law. Although that still leaves a lot of wiggle room for new work, it is much closer to an ultimate settlement of a topic without the need for additional work by the scientific community.

The methodological approach of a technological world seeks primarily to modify the potential of new knowledge and to acquire deeper understanding with the aim of exploring the potential of recreating the phenomena. This is much aligned to Heelan's (1997) approach/suggestion for a revisit and review of the natural sciences from the perspective of hermeneutic philosophy. In any case, the goal has been a clearer or at least a different assessment of the status of theoretical explanatory knowledge and its relation to the life world. Furthermore, some sense of how the current logical empiricist and the hermeneutic traditions relate to one another with respect to the short-term explanatory goals of science and the long-term goals of knowledge may possible be gained.

Hermeneutic theories, while emphasizing the collective and the relational, acknowledge the contribution of the particular individual as an active, performing initiator, albeit one who is also an agent of collective understanding. Evidence and claim must preserve the individual's contribution (Newell 1986), as placed within the potential knowledge level of nature (Anderson 2009).

Interpretations and Explanations

According to Vattimo (1989), Capurro (1995), the key in today's knowledge society, is our relation to what we do not know, in and through, what we believe we know. One of the major challenges of a scientific community is the constructed overcome of knowledge's partiality on phenomena, leading to a fully transparent and complete perspective upon a potentially chaotic and yet creative empiricism.

Creating a knowledge database initially needs a pre-define field of knowledge, which is usually dealt with under a classification in strict terminology of the field. That might actually be an objectivized pre-understanding collection of the phenomena descriptors, following specifically coded classes of data that can be interactively elaborated and enriched by the scientific community. That shall serve as an epistemic paradigm that conceives the information retrieval process more than just an interpretation process. The individual that has gathered information creates knowledge structures in order to actively interact with the system (Anderson 2009).

The appropriate philosophical approach to the method of interpretation is triggered by the breakdown of a task and begins by calling on the deep structure of pre-theoretical pre-categorical understanding of being which is found in the life-world. Inquiry is awakened when a directed question is formed. Which, like all directed questions, already implicitly contains an outline of a search and the discovery strategy aiming at uncovering a solution. The question construed in this case is not in an articulated form yet. Only later it achieves an adequate expression in what philosophers of science call an "explanation".

Philosophers have wondered whether science might be better off abandoning the pursuit of *explanation*. Duhem (1954), among other philosophers of science, claimed that "*explanatory knowledge* would have to be a kind of knowledge so exalted as to be forever beyond the reach of ordinary scientific inquiry: it would have to be knowledge of the essential natures of things. Something that neo-Kantians, empiricists, and practical men of science could all agree was neither possible nor perhaps even desirable" (Strevens 2006).

There follows an active dialogue and actions seeking practical fulfillment in the awareness that the sought-for understanding has presented itself and made itself manifest to the inquirer. If understanding is absent, search resumes, dipping again into the available resources. This hermeneutical circle of inquiry is repeated until a solution presents itself within a new cultural praxis in the life-world.

Within this book the cyclic development based on understanding of existing knowledge, seeking of missing one and progress through a repeatable process, will be extensively discussed in an attempt to provide a methodological approach for engineering the research in the most efficient and complete, in time, way.

Realizing Physical Phenomena

Gadamer (2004) had spoken of the *horizon* as "the range of vision that includes everything that can be seen from a particular vantage point. Applying this vision on the thinking mind, we talk about the narrowness of the horizon, or of its potential expansion, or the opening up of new horizons, and so forth." Following Gadamer's line of thought, Vamanu (2013) noted that the configuration of prejudgments, as it is related to a domain of objects, constitutes a 'horizon of understanding'. This horizon is a boundary for all possible phenomena we can experience either as the familiar ones or as those potentially cognizable. Furthermore, the horizon mainly constitutes a limit for understanding and a condition of possibility that is constantly shifting. Accordingly, we may recognize the aforementioned "horizon of understanding" as the *knowledge's technology potentiality* within the known world, prospectively to be increased and improved only via the optimization of its inherent property, i.e. the engineering of the knowledge. That may be achievable through a better understanding of the phenomena and the contribution of the combined reciprocity of theory and technology.

In developing this concept, we shall follow Seely's (1984) point, indicating that for facing the problem of adherence among the human behaviors and experimental procedures—considered to be typically scientific—we shall be potentially obstructing practical answers to engineering problems, while at the same time might have failed to improve the theoretical understanding of these problems. Thus, it is the collection and analysis of data via a repeated sequence of reproducing experimental "devices", an essential step to the apparent problems still to be identified.

The Essentials for Understanding

It is indeed both the data collection and analysis that unfold the experimentation method and process, connect the analogous opening of knowledge and, hence, leads us to the understanding of nature. This anticipating process constitutes the aforementioned *circle of understanding*, holding normative implications for research.

An iterated reciprocal movement between the meaning of a part and the meaning of a whole to which that part belongs under the assumption that a part only makes sense within a whole, may occur. Yet, within this *circle of understanding*, the whole does not make sense except in terms of a coherent configuration of its parts. This circle of understanding, contains the information-derived-knowledge, both information and knowledge constituting the expert system called the *hermeneutical cycle*. Therefore, the hermeneutical cycle, in its totality as a "device", depends on an inherent knowledge engineering functionality, which is transforming the existing knowledge, via appropriate justification means, into understanding: one of the knowledge technology's forms.

The collected data regarding the phenomena and thus, the development of knowledge, signify the integrated details within a phenomenon that shall provide a coherent and meaningful whole of the world of the phenomena. Every new finding has to be accepted and embedded in the pool of the existing experience (a potential "barrier") while, on the other hand, it may improve the value of this existing knowledge after its acceptance (a promising "advantage"). After being a part of the existing knowledge, each new discovery becomes one step beyond the subjection of the circle of understanding. The consequence regarding the understanding and design of experimentation systems, will be that in setting up a knowledge database, the fragmentation of information forces us to create the conditions of possibility for the retrieval of the knowledge pieces.

In order to move towards a precise reveal of an empirical-theoretical system, the experiment will have to satisfy three requirements:

(a) it must be *synthetic*, so that it may not represent a contradictory but rather a possible world;
(b) it must satisfy the criterion of *demarcation*, by not being metaphysical instead of representable of a world of possible experience and
(c) it must be *distinguished as a system* representing our world of experience compared to other similar systems to be submitted to tests, to which it has to stand up against (Popper 1959/2002).

A rather glaring point requiring further consideration concerns the well-defined experimenting "device", which needs to be engineered by the researchers well in advance and in pure methodological manner. When considering phenomena's ability to engage in the experimental "device", we move our thinking from a deterministic stance, towards a potential/probabilistic approach, by focusing on the classification of the systemic organization, through its categorical descriptors. This conception does not simply judge the phenomena's abilities, but moreover enhances understanding of how the environment influences the phenomena's progress, and the systemic capability to withstand the potential of disclaiming a hypothesis and ultimately pursue a specific roadmap.

The consequent usage of the operational profile of a system, justified through, among others, the physical and mathematical theories, the cognition on optimum alternative solutions, benefits, risks, factorization and analysis of means and targets, will lead to a methodology concept for an engineering based design of the systemic descriptors. A potential replacement of standing techniques by new ones, however obeying to comparatively different principles, allows the potentials, of those research community members that wish to reaffirm the substantial rationality of scientific approach and results, to advance.

The perception of a physical phenomenon, as an awareness perceived by scientists, is included within the first two rows of the above Table 1 namely "The World of Phenomena" and "Scientific Knowledge". The following two rows (namely, "Similarity of Knowledge" and "Classification of Knowledge") correspond to the compulsory step-wise procedures need to be satisfied for to guide the resulting decision regarding the experimentation outcome as presented in the last two rows (namely, "Experimental Design" and "Experimenting Engineering"). Furthermore, Table 1 is indicatively describing two modes of transition occurring in the experimentation phase, i.e. from experience to technology, through the classification of the existing knowledge, as well as from technology to technique, through the experimental design. The first transition is the starting point of the cognitional experimental design process, while the second one is supporting the practical experiment execution. The common operator in all of the above procedures is the human as a critical factor that perceives the phenomenon under

Table 1 A summary of a simplified three steps process for realizing physical phenomena, existing knowledge and experimentation

Steps	In	Process	Out
The world of phenomena	Categories	Hypothesis	Systems
Scientific knowledge	Measurements	Knowledge	Phenomena
Similarity of knowledge	Criteria	Similarity	Rules
Classification of knowledge	Values and cognition	Classification	A priori combinations
Experimental design	Conditions	Classes	Dissimilarities
Experimenting engineering	Mathematics	Experimentation	Values and parameters

investigation, categorizes the existing knowledge, identifies the lacks (i.e. the potential field for further but necessary research), detects the internal and external similarities of the phenomenon and, finally, integrates all of the above within the appropriate experimental design, before selecting any technological available means to potentially fulfill the aforementioned shortages.

The interaction of the above matrix with the human factor is bi-directional: understanding is the one way from matrix to humans while action is the reverse pathway from humans to matrix. Hence, understanding advances technology that leads to better techniques, so increases pure experience of the community (via scientist's accepted work), which feeds-back to the consequently following research.

An important note to be made here is that an experimental design should be adequately available (if and only if) the rows named "Similarity of Knowledge" and "Classification of Knowledge" are suitably filled. Evidently, the first two rows ("The World of Phenomena" and "Scientific Knowledge") must be filled first, as has been stated elsewhere (Kanavouras and Coutelieris 2017a). In the end, a fully, properly and acceptably filled matrix, of Table 1, may describe the adequate realization of a physical phenomenon under investigation, identify the potential areas of shortened information (i.e. the fields where the research should be directed) and guide the experimenters towards satisfying the research inquiries.

Apparently, several pathways might produce analogous results for specific questions. Nevertheless, Coutelieris and Kanavouras (2017) demonstrated that all of those potential matrices are similar, thus, producing equivalent outcomes, given a specific phenomenon. The cross-section of the matrices filled per scientific hypothesis, does not constitute a theory regarding this phenomenon. Since infinite matrices may be necessary to produce a new cross-section theory, each filled matrix (as seen in Table 1) either further supportively completes an existing theory or actively argues its validity.

Essentially, the proposed matrix of Table 1 highlights the way that the circle of understanding a phenomenon under question, runs. Every innovative observation, simulation result, idea of whatsoever, shall inevitably need proofs that can be justifiably obtained through a suitably filled matrix. Also, a methodological approach to define a specific and well-posed transition roadmap from the description of a system as far as the formation of a Prediction was described in detail by Coutelieris and Kanavouras (2016). According to the authors, in order to link experimentation to knowledge it is necessary to observe and realize the relative facts for to produce a survey of existing knowledge regarding the phenomenon and to identify the gaps along with the unnecessary repetitions within this knowledge survey. Thus, a classification scheme is necessary to capture every new finding. The scheme will essentially function as a knowledge pool where deriving knowledge is to be accepted and embedded within this pool of existing experience. The ultimate aim is the improvement of the existing knowledge value.

A consequence regarding the understanding and design of experimentation systems is that in setting up a knowledge database, the fragmentation of information forces us to create the conditions of possibility for the retrieval of the knowledge pieces. Observable incidents can be studied trough observations and/or mathe-matical simulations, rather focusing on the cohesions among the systemic quantities (variables, parameters, etc.) rather than on the quantities themselves.

The Cycle of Understanding

Given that the *cycle of understanding* is particularly referred to a specific phe-nomenon, Kanavouras and Coutelieris (2017b) showed that it is possible to methodologically approach the supportive evidential background of a theory, through a validation process of the theory-related phenomena. Such a process is based on their level of internal *similarity* (Coutelieris and Kanavouras 2017) and the consequent *similarity* of the conclusive macroscopic categorical descriptors named "outcomes". The elementary concept for that is to define the necessary criteria that are able to assure that similarity does exist. In that case, the world of specific phenomenon, as perceived by scientists, constitutes a four-dimensional vector space where each vector corresponds to a specific level of knowledge regarding the phenomenon under question. Given the mathematically well-defined pathway from the satisfaction of the similarity criteria to the estimation of the values for the parameters affecting the mathematical expression (equations) regarding the phe-nomenon, a linear mapping over this vector space expresses the internal similarity. In this context, each new cycle of understanding might put in question an existing theory and, consequently, generate a new one instead.

Popper (1959/2002) mentioned that a theoretical system has to be (a) free from contradiction (whether self-contradiction or mutual contradiction) this is equivalent to the demand that not every arbitrarily chosen statement is deducible from it, it must (b) independent, and i.e. it must not contain any axiom deducible from the remaining axioms. (In other words, a statement is to be called an axiom only if it is not deducible within the rest of the system.) These two conditions concern the axiom system as such. Regarding the relation of the axiom system to the bulk of the theory, it should be (c) sufficient for the deduction of all statements belonging to the theory which is to be axiomatized, and (d) it should be necessary, for the same purpose, which means that they should contain no superfluous assumptions. In a theory, thus axiomatized it is possible to investigate the mutual dependence of various parts of the system.

In order to tackle such systems, we encourage a proper handling of the "theories-technology-technique" triptych. It is this combination that will fully uncover and let us categorize the existing knowledge and reveal any potential gaps. Direct inter- and intra-developmental procedures among these areas of knowledge, as shown in Fig. 1, initiated developmental outcomes from each of the above sources of empiricism. Similarly, engineering has theoretical, technological or technical sources of origin within sequential developments (Mitcham 1999). This implies that the region where the three areas meet and overlap, consists the pure and applied zone for paradigms in each system (see Kanavouras and Coutelieris 2017b).

Fig. 1 The overlaying area
of the three major empiricist
contributors constitutes the
existing pure knowledge of
natural phenomena

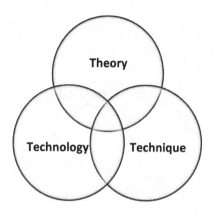

To fulfill such an approach, we propose to work on a cyclic mode, satisfying the intention that on every cycle each and every peripheral experience has the same equal distance from the "center". At the center of the cycle of experiences, the knowledge regarding each field may be positioned. Such a center could be named as the "dominating paradigm", which, according to Kuhn (1962) is usually established and commonly accepted, through years of empirical research and cognitional understanding. It is also constantly subjected to revisions, criticism and reconstructions until there are more problems rather than answers emerging from research in that field. Eventually, research will lead to a so called "scientific revolution" that through the new knowledge shall produce and establish a "new Paradigm".

Therefore, such a cyclic process shall now be described as an eight-step process, depicted in Fig. 2. The cycle contains the natural phenomena and a clear involvement of the experimenter in defining them. Each and every step of this cycle may be perceived as an individual process by itself, while each of these steps constitutes, in a sense, an individual "device" or a "gear", with a unique and particular functionality within the knowledge evolution process. Apparently, some of these steps rely mostly on the scholar knowledge available at a certain time period and practical experience (technical or technological), while others, although making use of accepted parameters, are mainly dependent on cognitional and individual interpretations of the researchers. Nonetheless, both of them will lead human decision making towards setting, defining and working on the research efficiency based on its outcome. Furthermore, experimental proficiency will now be reflected through the validation of the knowledge gaps (filled, remaining, or even opened), and the efficient use of resources for each particular research plan (Kanavouras and Coutelieris 2017b).

In brief, the cycle is triggered by the description of the empirically defined systemic categories which when placed into a certain context, shall form the hypothesis. Any hypothesis hence, needs to be presented through the relevant mathematical representations of the physical or chemical phenomena related to the system as well as the naturally developed sequences of phenomena occurring within

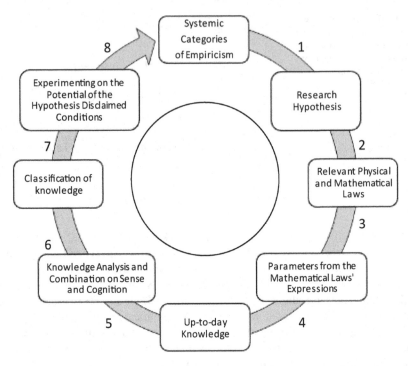

Fig. 2 The cycle of understanding the physical world and increasing knowledge of physical phenomena

an event, using the relevant mathematical description. Having established and identified such laws and relationships for the hypothesis, the available knowledge may then be analyzed and logically combined within the systemic and hypothesis boundaries, eventually leading to its proper classification. That, consequently, makes the experimenter capable of identifying the research gaps and selecting experimental conditions that will allow a proper expression of the relevant physical and natural parameters. Securing or disclaiming the hypothesis through this approach, will initiate an underlying new cycle for the newly refined hypothesis. This road-map is step-by-step presented and conclusively supported in Fig. 2.

In Fig. 2, each transition within the cycle represents a specific procedure towards the deep insight of the phenomenon under investigation. More precisely, for a given system, a falsifiable hypothesis must be defined (arrow 1) in order to be able to study the system under this hypothesis. To further proceed with understanding, it is now necessary to apply the relevant laws and principles (arrow 2) and to express them mathematically (arrow 3) before obtaining the results (arrow 4). After obtaining this information, it is necessary to obtain the cognition about the system (arrow 5) and to classify the existing knowledge (arrow 6) in order to identify the existing gaps and to design and perform new experiments (arrow 7), which will unavoidably define a new system and the cycling will now restart (arrow 8).

Consequently, in order to progress with the cycle's steps or validate and evaluate any research plan and objective, the classification of the existing knowledge is considered absolutely essential. This knowledge classification (step #6 of Fig. 2) asks for a satisfactory process that includes a thorough review of the available data and experience gained on the existences, along with a selection of both the systemic descriptors and their main participating classes. Then, knowledge could be captured within those classes-cells that have been mainly defined via the participants in the system.

In order to work according to this methodology, we need to implement the systemic characteristics, predicates, attributes, qualities or properties as part of the systemic experiences that control the evolution of the phenomena within a given system, placed at a particular environment. Furthermore, we also wish to propose a comprehensive methodology for fully meeting the research requirements. As a methodological consequence, conclusions of the research and studies should focus on determining any particular outcome in terms and conditions for their satisfactory fit to the given hypothesis. Moreover, the particular systemic participants, as recorded within the existing knowledge, need to be classified in order to point out their interrelationships (Kanavouras and Coutelieris 2017b).

For example, in order to approach the "food-packaging-environment system", we shall consider it as an engineering-based system following the "in-process-out" scheme. We may accordingly define "in" as being constituted of matter and energy, "process" as the relationships within and along the matter and energy, while "out" as the outcome perceived by senses. The scheme of such a system is given in Fig. 3. The main knowledge classes of this scheme will be further discussed in the following section.

In particular, matter, which in the case of packed food refers to all the materials constituting the system, consists of the following essential groups or knowledge classes that are actually developed with: (i) each and every experience, or empirical evidence, (ii) the properties, (iii) the qualities, and (iv) the characteristics, for any systemic possibility in general. Accordingly, "experience" may refer to the physical dimensions; "properties" of the matter can be related to the mass transfer phenomena among systemic materials; "qualities" in a sense adhere to the indicators, markers or factors relevant to the materials; and "characteristics" propose the molecular chemistry, internal composition, recipes and concentrations of interest to the hypothesis. Based on the above scheme, experimenters essentially modify the

Fig. 3 The four categorical descriptors of a system in the center and their main knowledge classes

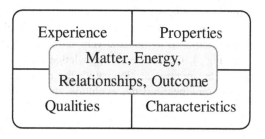

evolution "possibilities" of the system by critically utilizing knowledge, and then making cognitional decisions on how it justifiably complies with the hypothesis in question and, most importantly, the appropriateness of this knowledge. Having achieved that, innovative approaches (ideas), inventive applications (technology), practicalities (techniques) and/or alternative rationality (scientific theories) may emerge.

An assortment of contributing properties/attributes of packaging (materials, means and processes), that can be described in physical terms and measured within analytical capabilities, can potentially become model parameters for mathematical expressions. Hence, according to the aforementioned four categories and the knowledge classes within each one, a list of properties, specifications or attributes could be used to further elaborate on each and every group of "experiences", "properties", "qualities" and "characteristics" for the hypothesis, regarding packaging in particular.

For a "food-packaging-environment" system that may undergo spoilage and adulterations during its storage, the available "energy" is now discussed. This descriptor defines the progress of the phenomena within, as well as, among the elements of the matter. This descriptor may be further broken down into the experience that includes the system's energy requirements (empirically measured), the energy properties (may include energy transfer, storage, and/or systems capacity), the qualities (energy transformations within the boundaries of the system) and finally the characteristics (connecting the energy to its rate/coefficient, frequency, wavelengths and similar knowledge descriptors).

An inherent systemic uniqueness is reflected in the "relationships" among its contributors. For that, "relationships" may break down to the experience (sensed by the species reactivity), the affinity properties (of the species for each other), the qualities (such as the rate/coefficients, sequences, and/or levels of expressions) and characteristics (indicating the contributors' inter- and intra-dependencies, adequacies, deficiencies, imposing restrictions, catalytic performance etc.).

Frequently, "outcome" has been the one and only component of the collected data for several studies. In this class, the measured experience mainly referred to is quantity, the properties of outcome are adjacent to its identification, the qualities signify the selected indicator(s), marker(s) and significance, and the characteristics could be related to the molecular chemistry of evolved compounds.

Conclusively, geometry, volume content, area, permeation, transparency, pack integrity, food-packaging active/passive interactions, activation energy—thermal properties, energy transport, energy transfer, thermal capacity, amounts, polarity, active/passive interactions, active reactivity sites, sorption etc. are, in the opinion of the authors, essential considerations for packaging, carefully selected for the food consideration, as demonstratively summarized in Fig. 4.

In Fig. 4, the four main knowledge classes of the categorical descriptors of a system are presented, along with the conditions for essentially altering and controlling (engineering) the evolution "possibilities" of the system. In a sense, this

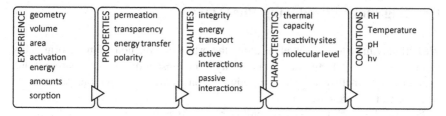

Fig. 4 Exemplary accomplishment of filling the four knowledge classes and the conditions for testing the hypothesis against them

last, fifth, column is the practical outcome of this method, and is of interest to the experimenters. These are the conditions under which the phenomena may be expressed or have an impact on the level of their expression. That research interest could be fully and clearly revealed via this approach, i.e. when the experimenters correlate the properties, specifications and attributes to the conditions that may impact on them. These conditions shall become appropriate for experimentation, directly linked to the measured empiricism and from that to the knowledge classes and systemic categories. Therefore, these conditions will check, within a minimum risk, the justified packaging that shall provide the least possibility for disclaiming the food preservation hypothesis (Kanavouras and Coutelieris 2017b).

B. Research in Practice

When at a starting point of a research attempt, the scientists apparently face a series of questions, dilemmas or even obstacles, they have to overcome in order to establish a research "protocol" they trust and can support. Let's recall some of these questions that at some point we had all been facing. For example, how may we proceed regarding the set-up and use of a particular research's scheme? Have we all the information for replicating an experiment? How can we trust the differences in tools and equipment, conditions and materials we are using in relation to those having been used by others? How may the experimental set-ups be compared among each other? Have we in the end wasted our research efforts and resources or have we actually achieved to look behind the closed curtains of the phenomena and reveal a hidden stage of physical activities, systemic performances, theoretical extensions or mismatches? How similar the findings or statements of one article, might be to those of another? Furthermore, how could we really comment on their similarity level and how may we select the one to implement against the other? i.e., when is something an exception or a contingency? Even if some of these have not been faced before, we believe we all agree are valid, important and may their answers may have a significant impact on the quality of any research and experiment. Therefore, in the following section of this book, we shall try to approach the answers, yet, not answering them in a unique way, which is left to each one with the interest to do so.

Creating and Reproducing Phenomena

We certainly have to ask ourselves about the origin of a phenomenon, its originality and uniqueness and furthermore, our ability to be present during its occurrence and evolution, or whether we may reproduce a phenomenon under our supervision within certain space and time.

One of the cardinal rules of philosophy is that one must always "save the phenomena". It is expected that despite of whatever a philosophical explanation might do, it should always account for the way things seem like to the spectator, to us. The principle called "saving the phenomena" is a powerful tool of criticism. The necessity of saving the phenomena is obvious for another reason, that is to say, when one considers the relationship between an "explanation" with the *explanandum*, as philosophers of science call technically *what an explanation explains*. Meanwhile, an aspect of a phenomenon apparent at one level might not be apparent to another. Since the ancient Greek times, philosophers like Plato (428/7–348/7 BC) and even earlier than him Parmenides of Elea (515–450 BC) they had saved the phenomena while claiming that the sensible world is illusory in a sense, because in their attempt to explain that phenomena are illusory they did explain those phenomena (Baggini and Fosl 2003).

The discussion regarding the creation of a phenomenon attracts an increased interest and becomes more meaningful in the case that a phenomenon may be reported, recorded, or even reproduced, though no solid theory is available to explain it, or embrace it within its establishments. Whether such a theory may ever be completed is an additional issue. Although, on the other hand, many phenomena were created after a theory had been established and accepted.

At the same time, there are theoretical entities which are used during the experiments in order to facilitate their functionality, support the conclusions and establish new theoretical relationships, via practical handling of such entities (see the entities used in modern physics regarding the structure of the matter). In other words, during the experimentation, we may use the theoretical entities, or even prepare them and handle under certain conditions based on their known properties. Additionally, we create new devices and structures also based on their properties, in such a way that, eventually, a specific outcome will be obtained. The aforementioned properties may have previously derived from pre-experiments and they may now be used as reasoned properties for the new experiments. In that sense, these properties are phenomena by themselves. Then, creating phenomena is based on the capability of isolating and use of these properties, as raw materials for creating and reproducing new phenomena. Then the theoretical entities are not the object of the experiment, but they rather constitute the mean to proceed with it cause otherwise the lack of materials will hinder the experiment.

Experimentation

According to Kant (1781/1998) experience deals with the raw materials provided by our sensible sensations. Thus, experience is rather the first product brought forth by our understanding. It is for this very reason the first teaching, and in its progress, it is so inexhaustible in new instruction that the chain of life in all future generations will never have any lack of new information that can be gathered on this terrain. Nevertheless, it is far from the only field to which our understanding can be restricted. It tells us, to be sure, what is, but never that it must necessarily be thus and not otherwise. For that very reason it gives us no true universality, and reason, which is so desirous of this kind of cognitions, is more stimulated than satisfied by it. Now such universal cognitions, which at the same time have the character of inner necessity, must be clear and certain for themselves, independently of experience; hence one calls them a priori cognitions: whereas that which is merely borrowed from experience is, as it is put, cognized only a posteriori, or empirically).

What is especially remarkable is that even among our experiences, cognitions are mixed in that they must have their origin a priori and that perhaps they serve only to establish connection among our representations of the senses. For, if one removes from our experiences everything that belongs to the senses, there still remain certain original concepts and the judgments generated from them, which must have arisen entirely a priori, therefore independently of experience. That could be said, because they make one able to say more about the objects that appear to the senses than mere experience would teach. Or at least make one believe that one can say this, and make assertions contain *subjective universality* and strict necessity, the likes of which merely empirical cognition can never afford.

Literature Search

The literature available for engineering sciences and technology has moved from predominant journal articles to a body of literature that also includes books, encyclopedia and conference proceedings, among others. Engineering science covers many aspects such as food, civil, mechanical, electrical, environmental, marine, etc. It is an applied science that also covers basic fields as chemistry, analysis, processes, based on a sound knowledge of the pure sciences including mathematics and biology. Engineers are also involved in developing engineering standards to promote and facilitate world trade.

Principal sources of information have emerged, highly specialized on topics and thematic fields, providing researchers with extensive and variable pieces of information and data along with critical reviews of highly appreciated topics. It is inevitable though that researcher may omit some information sources in any type of search that will mainly consider English-language sources, although in the case of local conferences, foreign-language materials may be included.

The Language of Science; the Terms of Truth

A critical point and consideration in research approach besides the language itself is the variability in terminology for the same or similar meanings. Such new worlds may signal new constructs, indicating lively science that is pushing against the boundaries of what is commonly used. It is *prima facie* plausible to postulate that there is nothing beyond understanding a *text*, that understanding the *sentences* composing it; and that there is nothing beyond understanding a sentence than understanding the *words* which compose it. The meaning of a complex expression is supposed to be fully determined by its structure and the meanings of its constituents (Szabó 2013).

Terminology is mostly used by authors and researchers for best describing their ideas, methods, tools and equipment and knowledge, although much less for providing findings, results and conclusions. It is not uncommon, particularly for young researchers, to have those new worlds trouble them in understanding their readings. To make things worse, messy nomenclature for the key-words that are frequently used as indicators of research and descriptors of the work may further confuse and irritate a researcher of the literature, while they may break down the bridges between one researcher's observations and another researchers' needs.

Language processing is a complex skill which has become routinized once one has gained experience in all important levels regarding understanding expressions: the phonologic, the semantic, the syntactic and the pragmatic. Over the course of time, sounds, words, sentences, and entire texts are automatically classified in one's cognitive system (Nehamas 1987) and therefore language processing takes place largely unconsciously under standard conditions. If a difficulty arises in the language comprehension process, and if one is not apt to immediately understand one or more linguistic expressions, then cognitive resources in the form of attention are activated, generating an interpretative hypothesis. This conscious process is often modeled as an interactive process of all relevant levels of information processing: the phonologic, the semantic, the syntactic, and the pragmatic.

The process of *parsing*, during which the words in a linguistic expression are transformed from a written sentence into a mental representation with a combined meaning of the words, as studied by cognitive scientists, is especially relevant. During this procedure, the meaning of a sentence is processed phrase-by-phrase and people tend to integrate both semantic and syntactic cues, in order to achieve an incremental understanding of a statement or a text (Pinker 1994). Yet alone, it was Wittgenstein who commented in his later work *Philosophische Undersuchunen*, on the meaningless meaning of a word as perceived by a human mind, but rather on the meaning obtained only through the actual use of this world, (Wittgenstein 1953).

A nexus of meaning, connected with a specific linguistic expression or a specific text, is construed by the author against the background of his goals, beliefs, and other mental states while interacting with his natural and social environment: such a construal of meaning is a complex process and involves both the conscious and unconscious use of symbols. Text interpretation can be conceptualized as the activity directed at correctly identifying the meaning of a text *by virtue of accurately reconstructing the nexus of meaning that has arisen in connection with that text* (Mantzavinos 2016).

As Rescher (1997) points out: "The crucial point, then, is that any text has an envisioning historical and cultural *context* and that the context of a text is itself not simply textual—not something that can be played out solely and wholly in the textual domain. This context of the texts that concern us constrains and limits the viable interpretations that these texts are able to bear. The process of *deconstruction* —of interpretatively dissolving any and every text into a plurality of supposedly merit-equivalent construction—can and should be offset by the process of *reconstruction* which calls for viewing texts within their larger contexts. After all, texts inevitably have a setting—historical, cultural, and authorial—on which their actual meaning is critically dependent".

What lies at the heart of this epistemic activity, i.e., of inventing interpretations as reconstructions of nexuses of meaning with respect to different aims, and how it can be best methodically captured is the subject of the application of the hypothetico-deductive method in the case of meaningful material as a plausible way to account for the epistemic activity of text interpretation (Føllesdal 1979; Tepe 2007).

Dealing with *specific* problems of interpretation, arising within specific disciplines like jurisprudence, theology and literature, which have been the focus of philosophical approaches. The aim was indeed to show what kind of *general* problems of interpretation are treated by the discipline of hermeneutics and to identify some important procedures leading to their efficacious solution—always keeping in mind that these procedures, like all epistemological procedures, are bound to remain fallible.

Hypothetico-deductive or *deductive nomological* method has been originally debated in connection with the philosophical theory of scientific explanation and can help establish hermeneutic objectivity, ultimately based on a critical discussion among the participants to the discourse on the appropriateness of different interpretations regarding the fulfillment of the diverse aims of interpretation. Inter-subjective intelligibility, testability with the use of evidence, rational argumentation and objectivity are, thus, feasible also in the case of text interpretation. A series of examples from diverse disciplines demonstrate this.

It has indeed been the case that the main protagonists, Hempel (January 8, 1905–November 9, 1997) and Popper (July 28, 1902–September 17, 1994) have portrayed scientific heart activity as exclusively an explanatory activity—largely aiming at answering "why?"—questions (Hempel 1962, 1948; Popper 1959/2002, 1963/2002). They both notably have argued that what lies at the heart of scientific enquiry is a starting hypothesis that proves e.g. that lead is heavier than water: if this true it is possible to deduce certain other claims, true ones that follow from it; one obvious is that lead sinks in the water! This influential and, very often, only implicitly shared view that all scientific activity is explanatory need not be followed, however. Moreover, answers to "what was the case?"—questions rather than only to "why?"—questions can be allowed to enter the field of science, appropriately accommodating the activities of all those whose daily work consists in text interpretation. The application of the hypothetico-deductive method is a way to show that the standards currently used when dealing with problems of explanation—intersubjective intelligibility, testability with the use of evidence, rational argumentation and objectivity—can also apply to problems of interpretation.

It is only the institutionalization of the possibility of criticism that can lead to the correction of errors when these evaluations and choices are involved. Our *fallible judgments* are all what we have here as elsewhere and *enabling a critical discussion* is the prerequisite of making informed choices (Mantzavinos 2016).

Information

The information-seeking process is basically a context-interpretation process of the people who store different kinds of knowledge in various conceptual formats. This process has a meaning within fixed contexts of understanding such as: thesauri, key words and classification schemes. Considering the process of storage and retrieval of information from the hermeneutical point of view, it could be considered as the articulation of the relationship between the existential world-openness of the inquirer, his/her different horizons of pre-understanding and the established horizon of the system.

The fragmentation of information forces us to create the conditions of possibility for the retrieval of the pieces because their common context remains tacitly implied. The partialization opens the possibility for different perspectives of interpretation. This situation can be described in terms similar to those used by Heidegger to analyze the structure of understanding: the general conceptual background (*Vorhabe*), the specific viewpoint (*Vorsicht*), and the corresponding terminology (*Vorgriff*) (Cappuro 2000).

Yet the system called 'empirical science' is intended to represent only one world: the 'real world' or the 'world of our experience'. In order for an empirical theoretical system to become a little more precise will have to satisfy three requirements, first, it must be *synthetic*, so that it may represent a non-contradictory, a possible world. Secondly, it must satisfy the *criterion of demarcation*, i.e. it must not be metaphysical, but must represent a world of possible experience. Thirdly, it must be *a system distinguished* in some way from other such systems as the one which represents our world of experience. The system that represents our world of experience when it has been submitted to tests, and has stood up to tests (Popper 1959/2002).

Knowledge obtained via experience is rather a public 'domain' where scientists can share the products of their understanding obtained through a well-defined experimental, observational and in-nature-reproducing, framework. It is implied that there have been developed common ways of working, mainstream action and terminology and communication conditions using scientific language and other media. How does the scientific knowledge interpreted by one scientist may by understood by another, if not via logical reasoning common goals, set up norms, definitions, similarities, gaps, criticism, testing questions, etc., aiming in the most efficient common knowledge and understanding, of all-times developments, transformations and innovations. Finally, though subject to change under transmission, it is not on this account devoid of truth, rather is meaning the instrument through which truthfulness makes its appearance in the life-world.

Hermeneutic cycle was conceived in terms of the mutual relationship between the text as a whole and its individual parts, or in terms of the relation between text and tradition. With Heidegger, however, the hermeneutic cycle refers to something completely different: the interplay between our self-understanding and our understanding of the world. The hermeneutic cycle is no longer perceived as a helpful procedural tool, but entails a cognitional task with which each of us is confronted. What matters Heidegger claims, is the attempt to enter the cycle in the right way. A suitable investigation into the ontological conditions of the system that ought to work back on the way in which universal cognition is led, is essential in the initiation of the cycle. What is needed is therefore not just an analysis of the way in which we de facto are trained by history but a set of quasi-transcendental principles of validity in terms of which the claims of the tradition may be subjected to evaluation.

Gadamer (February 11, 1900–March 13, 2002) recognized the importance of pre-judgments, which is a less loaded word in English than 'prejudices', in the *understanding* of language. For him, respect for authority and the traditions of one's community are paramount; *meaning* has to precede *understanding*. Habermas (June 18, 1929–) on the contrary, this being probably the most important aspect of his position and perhaps one that informs all his thought, believes that the project of cultural and philosophical modernity, which he distinguishes from *societal* modernity and the crisis in advanced capitalism, namely the working-out of the Enlightenment's enthronement of reason, is not dead—as the postmodernists in philosophy and in the arts assert—but is yet incomplete. For Habermas, *reason should be the basis of the consensus upon which communication depends*, not authority.

Hermeneutics, Habermas argues, must be completed by a critical theory of society. Following the above expressions of hermeneutics and attempting a transfer of the philosophical to technological sector, we may claim the parallel compliance to each other. It is in our view that the scientific evidences historically collected during time, needs an evaluation method in order to be validated by quasi-transcendental principles. Such a method may only be applicable within the classes of knowledge, rather than the individual knowledge points. It is only via the classification of knowledge that a critical approach to the scientific knowledge may be accomplished, since the criticism emerges directly through the systemic descriptors, at the three levels of possibility.

In addition to meanings seen on the basis of a common tradition of interpretation (with its presumption of historical continuity), legitimacy can be gained by other meanings independently of any presumption that there exists a historical continuity of meaning with the source through a common tradition of life, action, and interpretation. Each acting within its own horizon of research purposes is in dialogue with relevant data through its own empirical processes of testing and measurement (Heelan 1997).

Hermeneutic method is a processing interpretative work trying to provide a philosophical foundation for understanding how quantitative empirical methods give meaning to empirical contents, how theory-laden data depend on the self-assessed presentation (classification) of the measured entity as a public knowledge entity. Thus, via the knowledge classification, the hermeneutic framework will be capable of emphasizing the interpretative nature of the domain of objects of information science and of the researcher's access to it, the effects of the researcher's prejudgments on her choice of research topics and on her approach to them, the importance of iteration in the process of data collection and of integration of 'anomalous' details into a coherent, meaningful whole at the level of data analysis and the importance of repeating the study of the same phenomenon to acquire an ever richer understanding of it.

In particular, the double role of measurements with equipment assist in creating and refining both theoretical and technological meanings will be addressed as part of the reproduction of the natural phenomena within an experimental "device".

Nevertheless, the *information* is an opportunity for characteristic formations within the hidden dimension of language. The *information* can become a voice within the polyphonic nature of human *logos*, if and only if it is interrelated to the whole range of its hidden potentialities. If it is not, then we will have no more than an information society. The key issue in today's knowledge society is our relation to what we do not know in and through what we believe we know (Anderson 2009).

Data

Data collection and analysis, sets off the cycle of understanding. Further unfold of experimentation, connected to the parallel unfold of knowledge and hence, understanding of nature, constitutes the cycle of understanding anticipation, containing normative implications for research. For a hermeneutic *circularity*, the knowledge of the phenomena and thus the formation of knowledge, implies that it is the 'anomalous' details within a phenomenon that, when integrated, provide a coherent, meaningful entity of the world of phenomena. The apparent "anomalies" have still to be identified, collected and analyzed at the level of data collection, via a repeated sequence of reproducing experimental "devices". With hermeneutic theories in need of being empirically adequate, scientists should make significant and accountable choices for a human accomplishment within the semiotic domain. This truth construction accommodates the characteristics of the phenomenal world but is not determined by them (Schultz 1965; Weber 1974).

Whatever operations are performed for gathering information about people or objects of interest in a study, some trace of what is detected must have been captured and recorded. All the traces together are called "data". In natural sciences, data are coming from the empirical experiences of the researchers, either via direct or most of the times indirect means, tools and equipment. That is in particular, called raw data, even though if it is very frequently the outcome of a software or other type of treatment applied on the analysis experienced in regard to a phenomenon taking place during the experiment. A further process of the raw data may be required in order to closely approach, form or imply an answer to the initial question. It could as well be that the selected treatment of the raw data is selected in order to best fit and provide a pre-set opinion, point or objective of the researcher, concluding on a certainly biased outcome, remark and conclusion. Then, it is also possible that treated data requires an additional writing skill which as mentioned already could be expressed and interpreted on an individual base, yet alone that the individuals collecting the data may be different from those analyzing it and even writing about it.

In that process, the deep understanding and appreciation of the theory supporting and defining the analytical equipment, the technology impact on the handling and outcome of the resulted values and the human factor itself, need to be considered as significant factors of data collection and analysis. It is in total, an additional matter of trust and confidence regarding the manipulation and inspection of raw data to become processed, (most often statistically) analyzed and reveling. Furthermore, data rarely tells a simple univocal story, and this should not come as a surprise! With this in mind, the question regarding the quality of data collected and its truthful reflection of the phenomena intended to be examined, follow each and every single experimentation/data collection methodology.

Starting from these findings, one may wonder how to situate them in the context of existing knowledge outside the typological varieties of research studies' publications, but rather into a more scientific, natural laws-based and relationships-depended approach with a full inclusion of parameters, descriptors, conditions and outcomes. We may say so, since although there are many variations of research strategies that represent various ways of comforting a common set of problems, but besides the nomenclature differences, in the midst of variety, many familiar ideas may be encountered. Their application though, requires a certain type of expertise in strategic methodological classification of the data and knowledge as well as on the design of useful information extraction from that. Having a strong support to that saying, we wish to further propose and discuss a simple framework for categorizing studies, as well as proposing a conceptual basis for recognizing justifiable connections to further research planning and execution.

To cite Kant (1781/1998) again, for every concept there is *requisite*, first, the logical form of a concept (of thinking) in general, and then, second, the possibility of giving it an object to which it is to be related. Without this latter it has no sense, and is entirely empty of content, even though it may still contain the logical function for making a concept out of whatever sort of *data* there are.

Literature Briefs

It is apparently essential to search for and read the scientific publications, provided one has the experience via having received relevant education, training or guidelines for practical approaches, in doing so. The aim is to maximize the benefit of understanding, appreciating and ultimately utilizing their content, meanings, methods, examples, data, theories, opinions, ideas or even failures. All the aforementioned "benefits" contribute, to the formation of a hypothesis, or provide guidelines in formulating a methodology, understanding and appreciating, through compliance of the derived results to those previously produced by others.

It is of no surprise to mention that although an enormous number of articles have been sent for publications daily from various research institutes, groups of scientists and organizations, under numerous scientific programs, to many types of publishing journals of various prestige and acknowledgments, it still is a fraction of those selected as worth publishing. Within this framework, specialized editors are doing their best in managing the work load and subjectively handling the work coming on their tables, (needless to say) with a certain publication policy and a certain goal for the "best" possible selections.

So, we may use the world research as a term describing a multistage, multi-disciplinary process, with formal rules that usually describe each step, the first one being a careful defined question followed by the design of a systematic way of information collection and knowledge extraction. It is this process that may distinguish academic research from other casual types of searches for goods and daily news and information on a topic.

The perception of scientific research may be formalized *under six perceptions*, some widely held and other found among certain groups of readers, such as complexity of results, conflicting results, trivial topics, impractical studies, absence of commitment and care, and conflict with other sources of "truth". The first one, complexity, being a characteristic of both research-based knowledge structures and the way scholars think. Needless to say, that complexity is also the cause of apparent conflicts among research findings.

The following questions, selectively placed in three groups, is a simple, though by no means exhausted, attempt of organizing but also revealing the areas of importance and troublesome in reading, understanding and applying research results on the basis of the knowledge provided. These concerns by no means have the motive of discouraging researchers. On the contrary, the aim is to allow the researchers advance towards a realistic understanding of the status of a, sometimes, accorded research. It is also an attempt to bring together those topics that need to be considered in order to bridge the gap between worshiped and impractical research. The gap between like and do, believe and act, trust and apply. The gap to bring the researcher from being a follower to become an independent leader.

Engineering of the Experimentation

According to Popper (1959/2002), science can be considered by certain standpoints, and not only by an epistemological aspect; for example, one could perceive it as a biological or as a sociological phenomenon. As such it might be described as an implement comparable perhaps to some of our industrial equipment or even machinery. Science may be regarded as a means of production—as the last word in 'roundabout production'. Even from this point of view science is no more closely connected with 'our experience' than other instruments or means of production. And even if we look at it as gratifying our intellectual needs, its connection with our experiences does not differ in principle from that of any other objective structure. Admittedly it is not incorrect to say that science is '… an instrument' whose purpose is '…to predict from immediate or given experiences later experiences, and even as far as possible to control them'. Popper does not think that this talk about experiences contributes to clarity. He comments that it has hardly more point than, say, the not incorrect characterization of an oil derrick by the assertion that its purpose is to give us certain experiences: not oil, but rather the sight and smell of oil; not money, but rather the feeling of having money.

Experience deriving through experimentation with natural phenomena within an experimentation "device" performs as a distinctive method whereby one theoretical system may be distinguished from others. Hence, empirical science seems to be characterized not only by its logical form but, in addition, by its distinctive method. The theory of knowledge, whose task is the analysis of the method or procedure peculiar to empirical science, may accordingly be described as a theory of the empirical method—*a theory of what is usually called 'experience'* (Popper 1959/ 2002).

Engineers, on the other hand, prefer to see the world as subject to the laws and regulations of an ideally perfectly programmable, fully functional and strictly predetermined technical machine. A machine with complementary parts, comparatively and logically inseparable from competition, that is, a certain amount of "noise", i.e. "noise" is a certain amount of inconsistency of machines' operational status, produced by the interference of the contributing parts at their interfaces. Provided that this "situation" remains constant, this "noise" can urge the system into a new but mostly controlled status, a re-organization on a new basis, considering as inherent part of the system, all outcomes of the interferences within it. Engineers must keep the disorder of each system within strict limits and in regions with, artificially, low complexity of the system. Hyper-complexity that sometimes develops in a system corresponds to a qualitative "jump" of the system via enhancing certain features and weakening others. The new situation may have a positive impact on the world outside the planned system of the world with similar consequences for its controlling engineer. Thus, super-complex systems are looser hierarchically, less specialized, and not strictly centralized, largely dominated by strategies and skills, dependent on inter-relationships, and all subject to a non-controlled "situation" much alike the loss of muscle control (ataxia), noise and error.

Research Planning

As in any other type of planning, research requires decisions that facilitate the data collection, their *validity* and *credibility*. Validity, a major quality indicator for the scientific community, generally denotes the condition of being justified, therefore verified and accepted as "real" on the design and execution level of the scientific in regard to (i) the question (*hypothesis*) asked in the study (*internal validity*), as well as (ii) the validity of the results in present, future as well as in different places and by other scientists (*external validity*). Both validities are extremely important for engineers who are in favor of applications, technical approves and efficiency indicators-based qualifications.

Studies designed to resolve all possible threats to a consistent and true inference, although rather rarely found in the literature, ascribe significant importance on controlling validity, reliability and other elusive qualities present in a study, towards a certain degree of certainty in the success of an experiment. Considering that, any study is as close to producing reliably valid results, as the human skills and efforts can devise. A skillful research planner develops a plan that approaches the optimum and reaches asymptotically the best possible outcome, according to a predefined opinion. This describes the aim to maximize the significantly desired outcome, while minimizing the significant non-desirable ones.

What then makes sense to exist is a number of the typical procedures for decision-making, utilizing the optimization objectives and targeting the effort minimization at the lower cost possible. Procedures that initiate, develop and conclude dialectically with the available *raw materials* (inputs-outputs-humans) within particular circumstances. Implying that there is not a single solution to the issues, simply because the issues are not unique and completely repeatable after all. Yet the method is not a problem as it follows the suggested particular process, rather independent of the shape of the issues. Meaning that the engineering method itself should not allow for any space for multiple solutions to appear for a given problem, or otherwise this will classify the problem as "non-engineering" based one. The method selection among various alternatives that are based on comparing the input-output relationships, makes engineering design a distinct approach compared to other design forms.

The structure of such activities is considered to differ from the scientific methodology of knowing and learning, though at the same time it is in a strong analogy to that. Design is proposed as the unique method for practical activities, only differing fact objectives, targets—aims and potentially the raw materials used.

A final remark concerns the relationship of the engineering design to the self-restrictions implied, since engineers only deal with materialistic reality. Any imaginary conception can only be sourced from the rational natural physics and the latest development of mechanics. Furthermore, the engineering ideas are subjected to mathematical treatments, analysis and criticism, thus, much more far from the influence of the human senses.

The underestimated view of the experimentation significance by Aristotle changed in the 17th century, when Francis Bacon suggested that we should not simply observe the pure nature, but we also need to manage the world in order to reveal its secrets. The Scientific Revolution established this view and upgraded the experiment to the brilliant pathway to true knowledge. Nevertheless, the history of science is mainly the history of theories.

Once Bacon's essays and philosophies, regarding experimentation and observation, were firstly accepted and established, they soon enhanced people's desire to take advantage of them to harness nature for profit. Inevitably, the "study" and "research" of nature proved to be less about *changing* traditional attitudes and beliefs and more about *economy's stimulation.* By the middle of the following 18th century the Scientific Revolution gave birth to the Industrial Revolution that radically and dramatically transformed the daily lives of people around the Globe. Western society has been pushed forward on Bacon's model for the past three centuries. People often neglect the crucial and vital role *doubt* played in Bacon's philosophy. Even with powerful microscopes, there is still a lot that human senses miss.

Accounts such ones in this book, identify with considerations on experiments and research in general, and are registered in the sphere of intuitions, trying to get their "grip" on applied cases. Surely experiments have their own theory in the background with prejudgments and certain distance from theoretical considerations of science. Yet, observations, experimentation and analogy had been considered as essential features of basic sciences. Observation recognizes, reports and articulates the facts clearly and in details; analogy reveals the similarities among phenomena and experimentation, brings into light new facts, and thus, progressing knowledge. The observation, as it is guided by the analogy, is leading to experimentation, while the analogy that may be confirmed by the experiment, becomes a scientifically established truth.

The value of experiment has been consolidated since the 18th century, but the meaning of the experiment has been profoundly criticized in its perception by the scientists. The experiment was perceived as a mechanism that would bring a certain result and outcome, yet in some cases it was proposed only as a support to cognition which should be one step before the experiment, in order for the latter to have a meaningful contribution. It is the idea that must drive the experiment. Still in some cases experiments are performed simply by curiosity and may—or may not—be the initial stages of upcoming thorough investigations on a solid hypothesis base. That last point indicates a *circular approach* to knowledge with a variable starting point. There is no universal initiation step especially among the various scientific areas of interest and the epistemic level each area has progressed and developed so far. It goes without saying that a significant observation may be the starting point. On the contrary, it has been proposed that each purposeful experiment is dominated by theoretical embodiments. Of course, such observations cannot do or mean anything by themselves. In certain cases, phenomena enthusiastically provoke, but remain without further use or exploitation as nobody can see what they really mean, how they are correlated to something else or how they may be used. Some highly sophisticated in their conceptualization experiments have been conducted entirely through an accepted theory. Such experiments were unique therefore influential to the historic development of the particular scientific field, critical and most widely accepted by scientists as "worth trying", although their outcomes were not necessarily broadly accepted at the time.

At the same time, great theories were based on pre-experiments, yet others have fallen behind due to the luck of connection to the real world. Even some recorded phenomena remain useless without a solid theory. In special exultant occasions theories and experiments meet even though they are of different scientific fields' origins. That brings us to the particular example of astrology, where experimentation may not exist, but all are based on observations. In this case experiments may only be those mimicking other scientific fields, approximating the phenomena, yet, only after theory has completed a thorough approach giving birth to a well-shaped theory. Finally let us mention the shelf-existing experiments that may exist for long, independent of a proper theoretical assignation. These are experiments that may provide solid, accurate, novel, yet not particularly straightforward and easy to explain by the existing, at the time of execution, theories and believes. These were not rejected, nevertheless they were also not accepted by all.

According to Hacking (1983), in a broader picture, the experiment is the second part of a scientific activity next to theory. While theory is an attempt towards "*representing*", the experiment is the attempt to "*intervening*" the phenomena into the world. When experimenting, an attempt to control the phenomenon on question, via the available, well defined and hopefully well controlled conditions, is in focus. Therefore, we may say that every experiment actually guides the expression of the phenomenon according to the selected and established conditions within the lab, i.e. within a given artificial "world". While theory is based on speculations, cognitive pictures of an object, and a qualitative view of the world, it may have no direct relationship to the real world or the phenomena. The relationship among theory and experiment also defines the *scientific methodology* to be adopted. An additional activity establishes the connection between these views and the world, the activity of *calculation*. This activity refers to the mathematical transformation of an initial theoretical hypothesis in accordance to the real world. In that process, the experiment is connected via the calculation 'threshold' to the theory. Calculation is the intermediate step between theories and experiments. Furthermore, the creation of models may also take place. Models still remain indirectly dependent and indirectly compatible to either the theories or the experiments, while the phenomena may be described with phenomenological laws connected to these models. A synergy among theory, calculation and experiment prerequisites the acceptance of those theoretical entities for which the hypothesis will be intensively made.

By no means should we suggest that theory and experiments are clearly distinct entities and separated up to their merge. Their inner-relationship is obvious in experimentation, where each step has to deal with (i) a theoretical approach or a device, and (ii) an equipment or a system that by itself has been constructed, invented or built based on an internal theory that guided the technology and technique experimenters apply in the lab. Inventions are indeed the outcome of a process where theory and experiments are applied on a practical solution, although there are inventions that have preceded a theory. Such a case is the invention of the steam engine by Watt which was the evolutionary outcome of preexisting attempts well before the thermodynamics theoretical establishment. The experiments were the invention efforts for the technological advancement in the heart of the Industrial Revolution era. Thermodynamics has been a theory that organized, summarized, utilized, and extended the experimentation knowledge and its practical applications.

But do all the phenomena really exist in the world, longing for the proper experiment to reveal, describe and present them to scientists? In many cases the answer is negative. In physics, for example, phenomena are reproduced or developed in a lab under well-defined and controlled conditions that are, at the same time, technically produced through other analogously preceding phenomena. Such conditions allow the isolation, distinction and stabilization of the phenomena. And these are the characteristics that permit their reproducibility at lab scale. Even if we assume that the lab conditions may be reproduced in nature without the intervention of the experimenters that is only possible only after the whole preparation process has been fully completed.

Reproducing a phenomenon is not that simple and implies a series of activities including design, teaching and learning on how to execute an experiment. Most importantly it implies the knowledge of when does an experiment actually functions. Such skills and capabilities can be obtained in the lab (under and within certain human and social environment), making the recording of a phenomenon a rather obscure situation.

Inevitably, questions and debates arise regarding observations, picturing reality, critical experiments, measurement's accuracy and data values significance as well as the nature of experiments and the role of theory into them and the role of scientists overall. At the same time, the above points also imply an inherent and individual way of speaking about them. Is it the words and language also significant in talking about observations, facts and truth during an experimental observation or is it that each and every observation is embedded with a (specific?) theoretical consideration?

Yet alone, in certain cases, the ability to have a device capably functional for revealing the phenomena in a credible way is much more important than observation. Still the observer should be extremely competent in order to get the most out of a device or equipment used in data collection, especially on phenomenally paradoxical events that have been previously neglected or ignored by others. Such remarks may be a corner stone in the forthcoming work, but the upcoming experiments should and usually do, overcome simple observation. Training and education may have a part in developing an observation capability which nowadays goes well beyond the human senses.

It is not uncommon that experimenting scientists place observation on the top of their work, and the instruments, equipment and devices in the center of their ability to collect data. Consequently, both the above have the very first role in their conclusions, theoretical remarks and criticism. Thus, science progresses hand in hand with technology and techniques applied in the lab. We may still debate on whether there are indeed significant differences between observations and theory, but are aware of the strong impact our decision and adoption of a certain belief may have on our work. Such a case is making a decision on *which* things we may observe and whether that decision initiates from a theoretical impulse or an independent and pure interest.

Experimental Design

Although the objectives of quantitative and qualitative research are not mutually exclusive, their approaches to translating the world involve distinct research techniques and thus separate skill sets. Experience in quantitative methods is not required, but neither is it a disadvantage. Essential for our purposes, rather, is that all qualitative data collectors have a clear understanding of the differences between qualitative and quantitative research, in order to avoid confusing one with the other technique. Whatever a researcher's experience in either approach might be, a general grasp of the premises and objectives motivating each, helps develop and improve competence in the qualitative data collection techniques, detailed in this guide.

A theory explains why some event occurs (or does not occur) by providing a model of the causes or conditions that control its occurrence (or non-occurrence). Since an event is mainly consisted of the participating, structural, phenomena that follow the apparently well-defined natural laws, in order to explain the event, are therefore in focus and consequently, the goal of the model is the experimental prediction and control. Alternatively, a theory may explain a normality mode? Regularity among events by providing a model of the causes or conditions that, if fulfilled, necessitate the "regularity" noticed in these events. Theoretical questions that reach the researcher have to be formulated into sharp and accurate "technological" questions, in order to reproduce nature in the lab in order to reflect theory to the reality of the phenomena expression or vice versa. It is the experimenter, who has to formulate certain "technological devices", experimental equipment, through which a concrete answer to these questions has to be elicited. Of course, more questions may be pending or appear due and through to the process itself.

Concluding on whether a "technological" need is to be excluded, not necessarily in total and forever, goes via and through the questions to be answered. Such questions may also be part of a gradual implementation into the experimentation unfolding process. Apparently, such an implementation will have a rather obvious impact on the research subjectivity and its outcome.

Within the goal of a sensitive research, still associated with the need to face questions, it is a common aim to identify all possible sources of error, but at the same time try to avoid generalizations. Hence, researcher-selective theoretical approaches and structures dominate the design of the experimental work up to its execution, and even more up to a conclusive formalism in the laboratory, predominantly a theoretical one. Nevertheless, Seely (1984) concluded that the adherence to behaviors and experimental procedures, commonly considered to be typically scientific, has obstructed the development of practical answers to engineering problems, while at the same time failed to improve the theoretical understanding of these problems.

A consequence of the notion of *horizons of understanding* for research is that the researcher should be able to articulate the configuration of prejudgments bearing on the phenomenon, in awareness of the fact that complete self-transparency in this respect, is utopian. If there is a methodical dimension to understanding, it can only refer to the researcher's ability to make any potential outcome of understanding, as feasibly explicit and to possibly by keeping in check the erroneous prejudgments that block the researcher's access to the studied object. The presence of peculiarities in the data is a good indicator that we have not understood the phenomenon completely. A consequence regarding the understanding and design of experimentation systems is that in setting up, say, a bibliographic database, the fragmentation of information forces us to create the conditions of possibility for retrieving the pieces. We need conceptual backgrounds, for instance the scope of a data base, specific viewpoints (classification schemes), and terminology. The result is an objectivized or fixed pre-understanding. These backgrounds belong to historical, cultural and/or linguistic situations. There is no *knowledge in itself*.

One of the lessons of philosophical hermeneutics is exactly that; intellectual innovation of this sort depends on—indeed, is a manifestation of—the self-renewing power of tradition, of its dynamism, and its interpretability and reinterpretability. We can be open and have expectations with regard to the subject matter of a text because we belong to a culture or tradition (familial, educational, cultural, disciplinary, professional, organizational, etc.) which equips us with prejudgments and stereotypes and prejudice about anything that can become a matter of possible (research) interest. We can in principle understand only those phenomena with which we share a kind of meaning (Vamanu 2013).

Experimental Strategy and Experimental Results

Experimenting means reproducing, testing, improving, and standardizing phenomena. And phenomena are hard to be reproduced in a standard way. For this reason, we need to reproduce them, instead of simply discover or unmask them, leading to a laborious and time-consuming procedure, even more than one, actually. These procedures include design a functioning experiment, find out how to make an experiment work, and even more know when the experiment functions. Obviously, having these in mind, observing seems less significant in experimentation. What counts is the competency of districting the odd, the erroneous, the manipulative or the falsified environment, equipment and tool used in observing a phenomenon. It is then the equipment itself that should function rather than the experimenter who simply records the observations. For that, education, experience and practical thinking-designing- executing may contribute to unbiased results.

Repeating an experiment implies a way to value its repeatability and statistics have been extensively applied for that task. But how about the original experiment? Does it still maintain its validity versus the repetitions and can we actually repeat an experiment? Typically, repeating an experiment is an attempt of improving it against a more standard, less biased version of the phenomenon. While repeating involves similar or similarly functioning equipment, similar or similarly observing humans, similar or similarly functioning environment and so on, in some cases it troubleshoots the analysts on whether they should keep the results of a repetition or not. Apparently, it is the accuracy increase of a measurement aiming to exclude the systemic errors, instead of working on statistical means and standard deviations of a large number or data that may potentially accumulate systemic errors within less accurate or even erroneous results.

Engineering Based Design

Within the various Engineering Studies, the engineering design and methodology represent an essential part of the studies and experiences obtained through education. And although there is much discussion regarding the methodological approaches, it is the design methodology itself that distinguishes between engineering and scientific methodology.

In the end, engineering design is the effort to solve cognitively, using the available knowledge, the construction issues that shall provide the maximum economical and no time-consuming process of construction, construction itself, or even both. It is a systematic effort for minimizing effort. It starts as a cognition effort and ends as a physical effort. Everything in between may be called *design*. The cognitional effort during the design is not *knowledge* in itself (either scientific or engineering), but rather a pre-construction end-point when we need to conclude that this is the way to go.

Every design has to be abstract and provide the means for constructing, for bringing to reality, each of its conclusive points, even (at the same time) as potentially only one of the means for creating a more general model. The language behind the design, or for the design, is related to the functional relationships existing within the specific environment the design has been developed or the particular area it is meant to be applied to. Such examples include the process flow diagram for chemical engineers, the functional diagram for mechanical engineers or the circuit diagram for electrical engineers, as essential tools during the cognitional phase.

Lab

As mentioned above, the reproduction of a phenomenon has to take place within a well-defined frame that has given components under strict control of conditions. We shall call that a *device* which allows the transformation of the applied forces.

Devices have been designed and constructed for specific reason. In experimentation, such devices may be viewed as the extensions of human senses and human functions upon the natural phenomena and consequently the measuring electronic equipment as extensions of the human nervous system. The first part of the previous sentence is about a technical experience of a natural potential, while the second part is referring to a technical mean for the realization of an imaginary potential, in relationship to a broader discussion regarding the relationship between the human intuitive plausibility and the experimentation.

The insights gained in the lab can be extrapolated to the world beyond, and this is a critical maintained assumption underlying laboratory experiments; this principle is denoted as *generalizability*. Of course, several different formulations have been used to depict the relationship between the lab and the field, like *parallelism, external validity*, and *ecological validity*. Parallelism is traced to Shapley (1964) and it is said to be established if the results found in the laboratory hold in other, particularly real-world, situations under *ceteris paribus* conditions (Wilde 1981; Smith 1982). Campbell and Stanley (1963) introduced *external validity* as follows: "external validity asks the question of *generalizability*: To what populations, settings, treatment variables, and measurement variables can this effect be generalized?" Ecological validity has taken on a multifarious set of meanings, including the notion that a study is ecologically valid if "one can generalize from observed behavior in the laboratory to natural behavior in the world" (Schmuckler 2001). But, confusion arises because it is clear that Egon Brunswik coined the term *ecological validity* to indicate the degree of *correlation* between a proximal (e.g., retinal) cue and the distal (e.g., object) variable to which it is related (Brunswik 1955, 1956).

For that, a scientific experiment is not merely a magnifying lens. In sciences outside the social or political field, it usually needs to be independent of the human energy, although it needs a control or a guidance by humans, in the sense that humans (researchers, experimenters) "drive" the reproduction of the phenomenon, via the experimental set-up, to a certain direction, on a prefigured roadmap (with available or applied tools, means, utensils and apparatuses). The fundamental energy, the inherent driving force is originated *within the phenomenon*, the environmental factors and the participating matter through the naturally expressed relationships and processes. The experiment then should be and should remain independent of the human energy, for the phenomenon to be developed as is, not as we want it to be. Then the experimentation on a phenomenon seen as an object, is radically transformed from a static object to a carrier and a producer of functions or specific physical, chemical/electrical processes. The fundamental establishment within the experiment is the transition of the matter and energy to an output, through the device (set-up). Parenthetically, the measuring equipment is, in a sense, the set of tools provided to the phenomenon, as communication means among humans and nature.

The devices are closed systems that can be analyzed using kinetic or dynamic terms, to describe the combinations of *resistant* substances capable of evolving to certain directions and transact a work. The picture of the experiments may be described as closed evolution chains or as combinations of resilient parts naturally ordered to produce a given outcome. The quality of this work, or in other worlds its favorable or unfavorable outcome, defines the efficiency or effectiveness level of functionality for this device. Then, it is an engineer's role to change either the kinetics or the kinetics and the dynamics of the device (system), to deliver a particular outcome. The way the individual parts of this device is ordered consist the mechanism of the device. The way the input is becoming an output. The mechanisms within a device are to be considered at the initial step of setting up the device, but also as a hint in defining the case studies, hypothesis or proposals of and for a research.

The fundamental scope of Engineering is to describe the input-output transformations and to compare either the inputs or outputs with quantified terms (*quantities*) as much as for qualified ones (*types*). Therefore, devices must be distinguished from simple structures. The design and build-up of such experimental devices and the incorporation of the evolution related processes of a phenomenon in them, equals the availability not simply of a natural object but also of a procedure, as any particular phenomenon shall and should only take place within the device. The more distant this device remains from the human influence, the more objective the device's procedures with be, and thus, the independence of the phenomenon from the external uncontrolled influences shall remain. The *product* produced by the device is based on the phenomena that take place and the way they evolve therein. These phenomena, and their consequent *product*, will be further defined by the system's dynamics that are developed in time and in space of the device and as they were allowed, by the experimenter, the lab or the environment, to evolve. Hence, differences in tools, utensils, apparatuses even though they may seem alike, produce a different product when applied in small or larger scales of matter and energy. The transition from distinct scales to significantly differing ones has been a major engineering riddle, only to be enhanced by the complexity of the phenomena to be reproduced. Then, the phenomena, the procedures and the (natural) object making the procedures possible, have to be approached as total procedures, handled as devices and be controlled and guided as such. In accordance, the variability in the outcomes of such a device as well as the operators' impact will, after all, indicate the level of human control and/or manipulation applied to the device.

An interesting point concerns the fatigue of a device. This can be related to the exhaustion of the inherent energy, the de-structuring of the device's structure and the transformations towards a produced work, let alone the reduction in matter or the accumulation of "toxic" substances (by-products). It is in many instances the recording of the device's *operational life* that is indeed the scope of the research and its rate the actual experimentation outcome. Even more, the manipulative change of this rate may be the ultimate target for a scientist, who becomes an engineer of the phenomenon to guide its application to a much-desired direction. This target could be an additional reason towards setting and working for a continuous cyclic research evolution and developments in applied sciences.

Following the aforementioned approach, it is rather obvious that the experiment needs more than a researcher; it needs an engineer, a person capable of running the device in a well-defined operational way, in a well-established functional procedure and in a standardized perception. The phenomenon that is incorporated and reproduced in an experiment, becomes a technical system. Thus, it is less of an independent entity that may deliver an unpredictable number of different and distinct relationships and becomes a *moment* in a system of predefined relationships. The phenomenon (its relationships or processes) is transformed to a primary reality within which the matter (substances) are functioning as moments.

A final concern on experiments as devices for reproducing the phenomena, is the human factor and the role of the lab as a social-scientific environment where each and every expertise are developed, grown, established and propagate. Within that environment, as in each and every society, many rules have been established and are followed as working directives. Needless to mention the impact of such rules on the degrees of freedom in carrying out an experiment, performing a design, executing a protocol and elaborating on the results by the participating scientists. The connoisseurs and their expertise are in many cases well assigned among the members, restricting and imposing the working, handling/managing, performing and even understanding processes. It is quite interesting the way technologies reflect to their creators and users, affecting their self-images, self-understanding and self-interpretations. Then experiments are who we are and vice versa. In conclusion, the phenomena and their devices cannot be independent of their frame, namely, their systemic world and its social interchange with the experimenters, the totality of which we call the meaning of the phenomena, or simply the experiment.

Equipment

Instrumentation during research has a major role in the method as well as the outcome of any experimentation set-up. In certain cases, the quality and quantity of the data collected is closely connected to the consistency of the collection process, bringing up the *reliability*, both of instrument and user mode.

Calibration and checking is an essential pre-step in the experimentation process, well in advance of the actual usage and application of the equipment of the study itself. Manuals, guidelines and procedures may be available, or even developed by researchers as part of the experiment set-up. In some cases, due to the close impact on the experimenters' experiences and empirical connection to the phenomenon under investigation, they need to be clearly mentioned in literature as an inherent part of the experimental data collection, analysis and validation. That may allow for another group to replicate the experiment, and/or correct, improve or criticize its results.

It is a great challenge for both scholars and experimenters to have a critical approach to different results among replicates, in combination to the conditions applied and the control means used to check the human interference and the non-predicted parameters. That in the top of its application may allow them not just to fit the results in the existing theory or practice, but going one step further. That is, to extend and expand the areas of understanding to the complex fields of unpredicted-yet well-known and defined-peculiarities of the experimentation process overall. After all, it is the *mistakes* or *faults* that have led to different observations and consequent new discoveries in sciences and technology.

Nevertheless, the *analytical* and *combinatorial* ability of the Experiment and scholar to reveal such opportunities and take advantage of the understanding of the relevant factors behind the *incomprehensible* phenomena, shall allow for an *engineering* of the phenomena and a creation of the new experimenting machine to function towards a different, but desirable, *outcome*. Both the skill of analysis and knowledge combination in support of understanding the world seem of paramount importance in planning, executing and validating an experiment. This is realized only when it is grounded on sound knowledge, not simply in its totality but in each and every one of its classes, participating and contributing to the lab scale reproduction of the world.

An important dilemma though, regards the point in time for confidence, reliability, analysis and combination, needed for the execution and understanding of the phenomena and the valuable answer to the question (hypothesis). In general, it is the particular form of research that could define whether the above confirmations may be established in advance, simply due to the amount of data to be collected and the human versus instrumental intervention in doing so. In the case of basic, applied and engineering based experiments it is great confidence to the instruments customarily accepted by the specific scientific community, leaving back the human factor, while the amount of data (or its collection and recording frequency) is usually considered adequate in the light of operator's experience, execution consistency, and time, cost and equipment availability. Numerous publications have been occasionally available due to a newly developed and used analytical device, creating new types of data collection, claiming high accuracy and maximum sensitivity in details recording, mainly due to recent advancements in the technology of the device, rather than the theory behind the analysis of involved phenomena.

Tools

Considering every equipment as a *machine*, or in other words as a structural totality of individual parts brought together in a certain operational and functional manner for performing a certain task of actions with an accepted performance efficiency, accuracy, reputability and reproducibility, and with operation input and recordable outcome, it is easy to conclude on the importance of tools to operate such a machine. Not just internal but external tools are involved, including calibration tools (machines by themselves), maintenance tools, means and methods, people (*human machines* with operational functionalities, skills and capabilities) or even environmental conditions that influence or collaborate with the equipment to allow its intended functionality (for instance, a thermometer has to be exposed to a *thermo-environment* in order to function).

Efficiency, Effectiveness, Economy

In defining design for engineers, we may mention the inherent intention to fulfill a construction, i.e. a reified intention. During this process, the design includes a process that incorporates the well-known techniques and technologies as well the scientific principles to define a cognitional object in such a way that this object can be naturally materialized as intended to. It is also a repetitive process of decision making for such designs that may utilize the available resources in the most optimum ways, thus the human mind is invited to fulfill the needs in the best possible way.

But besides the optimization, it is also the *fit-for-purpose* or *bounded rationality* that is included in the design process and targets. The optimization logic may distinguish among the *command variables*, means or entries, the *fixed parameters*, rules or principles and finally the *constrains*, targets or outcomes, but the process is aiming in fact in defining those values of the above parameters that will provide a widely-accepted sum. A sum should then be compatible to constrains, while at the same time it should also be maximizing the utility function within a given environment or conditional background. Given the complexity of the real situations, is good enough to have a design method that shall allow the selection among x potential alternative solutions, for the aforementioned *fit-for-purpose* or *bounded rationality*, to be met at the highest possible level, or at least provide a solution that will be better than the existing ones.

Results

The interpretation of the derived results for a system may be also expressed as a potentially alternative systemic model. Hence, the outcome will be a system of analytic statements (since it will be true by agreement). Such an interpretation cannot therefore be regarded as empirical or scientific, since it cannot be disproved by the falsification of its consequences, for these too, must be analytic. For a system to be interpreted as an empirical or of a scientific hypothesis, the primitive terms occurring in the system are to be regarded as 'extra-logical constants'. In this way, the statements of the system become statements about empirical objects, that is to say *synthetic* statements. This leads, however, to difficulties because a definite empirical meaning is assigned to the experimentation concept by correlating it with certain objects belonging to the real world, experimentally determined. Thus, the experimentation concept can be regarded as a symbol of those objects.

As Popper notes (1959/2002) "it is usually possible for the primitive concepts of an axiomatic system such as geometry to be correlated with, or interpreted by, the concepts of another system, e.g. physics. This possibility is particularly important when, in the course of the evolution of a science, one system of statements is being explained by means of a new—a more general—system of hypotheses which permits the deduction not only of statements belonging to the first system, but also of statements belonging to other systems. In such cases, it may be possible to define the fundamental concepts of the new system with the help of concepts which were originally used in some of the old systems".

If the hermeneutical circle describes the structure of understanding, the 'fusion of horizons' represents its mode (Gadamer 2004). Understanding emerges as an event out of this sort of fusion; the interpreter's horizon has expanded and has been enriched as a result of this 'merger', by acquiring a wider and more sophisticated view of the phenomenon. The event in which 'the tension' between the separate horizons of the researcher and of the object is 'dissolved,' as they 'merge with each other' (Gadamer 1989), can be achieved via the knowledge classification (see Vamanu 2013).

Thus, when we deal with research objects (especially connected to the past), prejudgments about these objects derive from a 'chain of interpretations' which have accumulate to become an integral part of the objects and thus their knowledge as they mediate our access and relation to these objects. In this respect, new questions, new concerns, and new contexts of research enable different under-standings of the same object and transform understanding into an unending endeavor.

A chain of logical reasoning for empirical scientists, can be validated, according to Popper (1959/2002), only when it has been broken up into many small steps, each easy to be checked via mathematical or logical techniques of transforming sentences. Raised doubts can only be pointed out as errors in the steps of the proof, or re-thinking of the matter. Describing experiments present empirical scientific statements that can be tested by skillful experimenters.

Engineering

A world of Latin origin *ingeniare* mid-14c., *enginour*, "constructor of military engines," from Old French *engigneor* "engineer, architect, maker of war-engines; schemer" (12c.), from Late Latin *ingeniare* (see engine); general sense of "inventor, designer" is recorded from early 15c.; civil sense, in reference to public works, is recorded from c. 1600, but not the common meaning of the word until 19c (hence lingering distinction as civil engineer). Meaning "locomotive driver" is first attested 1832, American English. A "maker of engines" in ancient Greece was a *mekhanopoios*.

Although great changes in the meaning appeared in Europe or USA, today it refers to the art of applying science for the optimum use of natural resources to the humans' benefit. It is the conception and execution of a plan for a construction or a system that functions under and responds to certain conditions at the most optimum possible way. In a sense, it is a cognitional study process for designing a device or a system that shall solve a problem effectively or shall confront a certain necessity. Then, an engineer should go beyond the actual making, fabricating or constructing, to managing, designing or studying in a systematic way. The engineer has to establish and order the particular engineering framework that puts all elements together.

Furthermore, engineering science, i.e. the systematic knowledge of making and relevant engineering crafts, contains the traditional as well as the latest upcoming branches that have particular application targets. Therein it is found the distinction among engineering and technology, as the first one dealing with the broader problems of application, while the later dealing with the more specific and particular issues. Needless to say, that the term may be used as either a mechanical-technical or a social sciences signification. Yet, both promote the conceptual creation of materialistic artifacts and many elements and influences that interact within this primary procedure and derive from that, influenced by its different forms and ultimately affecting them, in turn. In either case, such a creativity process is highly guided by modern science giving birth to a series of relevant references.

C. Mathematics

Mathematics and physics are the two theoretical cognitions of reason that are supposed to determine their objects a priori, the former entirely purely, the latter at least in part purely but also following the standards of sources of cognition other than pure reason. According to Kant (1781/1998) since the early time of ancient Greek philosophers, Mathematics has travelled the sound path of a self-sustained science. Yet, it was not as easy as it was for logic—in which reason has to do only with itself—to trace that quasi-royal path, or rather itself to pioneer it; rather, he claims that Mathematics was left groping about for a long time (chiefly among the Egyptians), and that its transformation is to be ascribed to a revolution, brought about by the happy inspiration of a single man in an attempt from which the road to be taken onward could no longer be missed, and the secure course of a science was entered on and prescribed for all time and to an infinite extent. The history of this revolution in the way of thinking—which was far more important than the discovery of the way around the famous Cape II—and of the fortunate ones, who brought it about, has not been preserved for us. But the legend handed down to us by Diogenes Laertius—who names the reputed inventor of the smallest elements of geometrical demonstrations, even of those that, according to common judgment, stand in no need of proof—proves that the memory of the alteration wrought by the discovery of this new path in its earliest footsteps must have seemed exceedingly important to mathematicians, and was thereby rendered unforgettable. A new light broke upon the first person who demonstrated the *isosceles* a triangle, whether he was called "Thales" or had some other name. For he found that what he had to do was not to trace what he saw in this figure, or even trace its mere concept, and read off, as it were, from the properties of the figure; but rather that he had to produce the latter from what he himself thought into the object and presented (through construction) according to a priori concepts, and that in order to know something securely a priori he had to ascribe to the thing nothing except what followed necessarily from what he himself had put into it in accordance with its concept.

Mathematical Modeling

Modeling is a scientific field where the use of mathematics is inevitable. As discussed above, mathematics actually offers a closed structure (language) to describe physical problems and to study them through the solutions (mathematical entities such as functions, along with several operators over them) obtained by this mathematical description. To model a phenomenon, it is necessary to separate the unknowns/variables from the parameters affecting them, to describe the processes with operators and tensors acting on the variable in several manners so that to (i) produce equations, (ii) to superimpose the conditions on the boundaries in order to sufficiently describe the singularities possibly being there and, finally, (iii) to use the appropriate mathematical techniques/methods in order to solve the mathematical equations, thus obtaining a set of mathematical entities (usually functions) that represent the consequent physical entities under investigation. The application of this solution on a variety of similar problems allows for the transition of its value from the world of mathematics to the real world of physical phenomena. The following graph clarifies further this recursive relationship between "nature" and mathematics (Fig. 5).

Two main approaches are encountered in the modeling world: the *deterministic* modeling and the *stochastic* one. To produce a deterministic model, one should apply fundamental principles and laws in the field where the processes under investigation occur, and therefore, to describe mathematically these principles in a form of specific terms containing ratios of variability (derivatives) for the quantities under question. By further applying balances, these terms form derivative equations along with conditions applied on the boundaries of the domain in which processes are supposed to take place. In fact, the produced system of equations is independent from the phenomenon itself that has to be described, since the mathematical theories used for achieving a solution actually consist applications (i.e. methods) based on broader and deeper mathematical theories.

The stochastic modeling is quite analogous to the thermodynamics approach for nature. In accordance to this approach, each medium is represented by a finite number of elements that have pre-specified degrees of freedom. At each time step,

Fig. 5 The correspondence of mathematics to the physical world

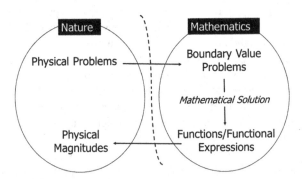

every element alters its situation (position, momentum, etc.), "selecting" one of some possible new situations/values under the limitation of a pre-defined possibility. Although the variability in a microscopic scale is absolutely randomly defined, the macroscopic behavior of the system that is described by statistically derived quantities such as mean, deviation, etc., is usually in excellent agreement with the expectations coming from the deterministic view of the simulated system. In that sense, this type of simulation is compatible to the opinion established by the thermo-dynamical approach of the phenomena: even though on *particles* level the activity is much random and disordered, when the various essential particles become a unity as a total (i.e. when we consider a great number of them as one, along with their interactions and interrelationships), then the system is *statistically* (inductively) predictable.

Apparently, both approaches produce solutions that are depended not only on the parameters affecting the physical system but also on mathematical quantities that are intrinsic properties of the theories, methods and tools used in the particular simulation. This cohesion between nature and mathematics is beyond the description abilities that mathematics offer as a well-structure universal language. It can be attributed in the "nature of mathematics", that is their inherent capability to keep a distance from the phenomenon into which they appear. The involvement of mathematics in description of physical problems allows for applications in terms of engineering, since engineering may be considered as the application of the *cycle of understanding* on the sensory world, i.e. what is perceived via the senses in combination to the available knowledge background.

Speculations, Calculations, Models and Approximations

Pragmatic science has a structure of its own, following a certain articulation of a theory that is then exposed to experimentation. The initial *speculations* may not be related to the real world, while, in many recorded cases, verification required new methods and technologies in the experimentation process. Thus, both theory and experiment need to be articulated accordingly. A theoretical approach to either one may be called *calculation*, i.e. a mathematical version of a given speculation for better fit to the world. It is the speculation that aims to a qualified structure of a scientific field, while the experimentation targets its own existence within this field. What the calculation process does is the bridging between those two, by a hypothetico-deductive model. This connection allows for a quantitative agreement between theory and experiment. Somewhere in between these steps the models for the phenomena and models for the theories, simple mathematical expressions and footprints, and representations of the world appear. The human senses and cognition assisted by advanced technological tools turn these models feasible. The consequence of such an action is the creation of an impression for the phenomena, the theories and their interconnection via simplified mathematical propositions. Within this frame, the phenomena become a reality and theories are approaching the truth more and more closely. The type, complexity and number of correlated or independent models may be used within the same theory as the only accepted representations of the phenomena. In some cases, models were proven more resilient than theory, since there is more truth to incompatible models than to a given sophisticated theory.

There is a tremendous number of phenomena, but only a few relative simple laws that are applied in nature. However, not all of these laws may be applied in all phenomena. The trend is towards the successful use of more incompatible models for the phenomena in every day's work than before, so the ultimate end-point shall be the absolute plethora of models and a unified theory. But even so, the vast majority of science and technology will be unattached, as there is a sustainable need for applications, as they develop per case. For that, engineering should be established on the comprehensible bases of theories, models, phenomena and proofs allowing for well-justified approaches to practical solutions. An engineering based justified application for manipulating and managing the phenomena has an additional circular effect on the origination of the models and the understanding of the phenomena, with a potential impact on the theories as well, through a cycle of understanding and re-organizing of our sensorial experiences and cognitional process.

Now, as Kant wonders (1781/1998) "why is it that here the secure path of science still could not be found? Is it perhaps impossible? Why then has nature afflicted our reason with the restless striving for such a path, as if it were one of reason's most important occupations? Still more, how little cause have we to place trust in our reason if in one of the most important parts of our desire for knowledge it does not merely forsake us but even entices us with delusions and in the end, betrays us! Or if the path has merely eluded us so far, what indications may we use that might lead us to hope that in renewed attempts we will be luckier than those who have gone before us?"

We should think that the examples of mathematics and natural science, which have become what they are now through a revolution that brought about all at once, were remarkable enough. This is why we could reflect on the essential element constituting the change in the ways of thinking that has been so advantageous to them, and, at least experimentally, imitate it insofar as their analogy with metaphysics (rational cognition) might permit.

Underlying the *literalist* view there are a series of cropped understandings (misunderstandings?) of the nature and role of mathematical models. This concerns, for example, how theories relate to the empirical world, the nature of truth, and particularly, knowledge as only a short causal objective snapshot in contrast with knowledge as long term dynamic, historical, and social assessments that function of necessity in a cultural milieu. The latter, being praxis-laden, does not need or support unlimited iniquity, precision, or causality (see Heelan 1997).

Finally, according to Popper (1959/2002) the admissible values of the 'unknowns' (or variables) which appear in a system of equations are in some way or another determined by it. Even if the system of equations does not suffice for a unique solution, it does not allow every conceivable combination of values to be substituted for the 'unknowns' (variables). Rather, the system of equations characterizes certain combinations of values or value systems as admissible, and others as inadmissible; it distinguishes the class of admissible value systems from the class of inadmissible value systems. Correspondingly, systems of concepts can be distinguished as admissible or as inadmissible by means of what might be called a 'statement-equation'. A statement-equation is obtained from a propositional function or statement-function; this is an incomplete statement, in which one or more 'blanks' occur. Now what Popper calls a 'statement equation' is obtained if we decide, with respect to some statement function, to admit only such values for substitution as turn this function into a true statement. By means of this statement-equation a definite class of admissible value-systems is defined, namely the class of those which satisfy it.

References

Anderson JA. Thinking qualitatively: hermeneutics in science. In: Stacks DW, Salwen MB, editors. An integrated approach to communication theory and research. 2nd ed. Routledge: Taylor & Francis; 2009.

Baggini J, Fosl P. The philosopher's toolkit. A compendium of philosophical concepts and methods. Malden, MA: Blackwell; 2003.

Campbell DT, Stanley JC. Experimental and quasi-experimental designs for research. Boston: Houghton Mifflin; 1963.

Brunswik E. Representative design and probabilistic theory in a functional psychology. Psychol Rev. 1955;62(3):193–217.

Brunswik E. Perception and the representative design of psychological experiments. 2nd ed. Berkeley, CA: University of California Press; 1956.

Capurro R. Moral issues in information science. J Info Sci. 1985;11(3):113–23.

Capurro R. Die Informatik und das hermeneutische Forschungsprogramm. Informatik Spektrum. 1987;10(6):329–33.

Capurro R. Leben im Informationszeitalter. Berlin: Akademie; 1995.

Capurro R. Hermeneutics and the phenomenon of information. In: Mitcham C, editor. "Metaphysics, epistemology, and technology", research in philosophy and technology, vol. 19. San Jose/Amsterdam: JAI/Elsevier Inc; 2000.

Chick G. Writing culture reliably: the analysis of high concordance codes. Ethnology. 2000;39 (4):365–93.

Coutelieris FA, Kanavouras A. A methodological approach on experimentation engineering. Experiment. 2016;36:2246–57.

Coutelieris FA, Kanavouras A. On the mathematics about similarity of physical phenomena. J Num Anal Ind App Math. 2017. (accepted)

Denzin N, Lincoln Y. Introduction—entering the field of qualitative research. In: Denzin N, Lincoln Y, editors. Handbook of qualitative research. Thousand Oaks, CA: Sage; 1994.

Duhem P. The aim and structure of physical theory. Translated 2nd edition La Théorie Physique: Son Objet, Sa Structure 1914 Paris. Princeton: Princeton University Press; 1954.

Froehlich TJ. Relevance reconsidered—towards an agenda for the 21st century: introduction to special topic issue on relevance research. J Am Soc Inf Sci. 1994;45(3):124–34.

Føllesdal D. Hermeneutics and the hypothetico-deductive method. Dialectica. 1979;33(3–4):319–36.

Gadamer, HG. Text and interpretation. In: Michelfelder DP, Palmer RE, editors. Dialogue and deconstruction: the Gadamer-Derrida encounter. Albany, NY: State University of New York Press; 1989.

Gadamer HG. What is truth? In: Brice R, editor. Hermeneutics and truth. Evanston, IL: Northwestern University Press; 1994.

Gadamer HG. Truth and method, 2nd rev. ed. London, UK: Continuum International Publishing Group; 2004.

Gödel KF. Über formal unentscheidbare Sätze der 'Principia Mathematica' und verwandter Systeme, I. Monatshefte für Mathematik und Physik. 1931;38:173–98.

Goffman E. The presentation of self in everyday life. Garden City, NY: Doubleday; 1959.

Hacking I. Representing and intervening: introductory topics in the philosophy of natural science. Cambridge, UK: Cambridge University Press; 1983.

Heelan PA. After post-modernism: the scope of hermeneutics in natural science (1997). Available at: http://www.focusing.org/apm_papers/heelan.html.

Hempel CG, Oppenheim P. Studies in the logic of explanation. Phil Sci. 1948;15(2):135–75.

Hempel CG. Deductive-nomological vs. statistical explanation. Minneapolis: University of Minnesota Press; 1962.

Kanavouras A, Coutelieris FA. Systematic transition from description to prediction for the oxidation in packaged olive oil. Food Chem. 2017a;229:820–7.

Kanavouras A, Coutelieris FA. A methodological approach for optimum preservation results: the packaging paradigm. Int J Food Stud. 2017b;6:56–66.

Kanavouras A, Coutelieris FA. Addressing theoretical & practical aspects on knowledge classification for engineering tasks; 2017c. (submitted)

Kant I. The critique of pure reason. Translated by Guyer P, Wood AW. Cambridge: Cambridge University Press; 1781/1998.

Kobayashi A. Ten years on: is self-reflexivity enough? Gend Place Culture. 2003;10(4):345–9.

Kuhn TS. The structure of scientific revolutions. Chicago: University of Chicago Press; 1962.

Lancaster K. Variety, equity and efficiency: product variety in an industrial society. New York: Columbia University Press; 1979.

Mack N, Woodsong C, McQueen KM, Guest G, Namey E. Qualitative research methods: a data collector's field guide. NC: Family Health International; 2005.

Mantzavinos C. Hermeneutics. In: Stanford encyclopedia of philosophy. Published online 22 June 2016; 2016.

Miller DL, editor. The individual and the social self: unpublished work of George Herbert Mead. Chicago: University of Chicago Press; 1982.

Mitcham C. Thinking through technology: the path between engineering and philosophy. Chicago: University of Chicago Press; 1999.

Nehamas A. Writer, text, work, author. In: Cascardi AJ, editor. Literature and the question of philosophy. Baltimore: John Hopkins University Press; 1987.

Newell RW. Objectivity, empiricism, and truth. In: Paul K, editor. London: Routledge; 1986.

Peirce CS. The collected papers, vols. 1–6. In: Hartshorne C, Weiss P, edirors. Cambridge MA: Harvard University Press; 1931–1936.

Peirce CS. The collected papers, vols. 7 and 8. In: Burks A, editor. Cambridge MA: Harvard University Press; 1958.

Pinker S. The language instinct. New York: Perennial Classics; 1994.

Popper K. Conjectures and refutations. London and New York: Routledge Classics; 1963/2002.

Popper K. The logic of scientific discovery. Taylor & Francis e-Library; 1959/2002.

Popper KR, Eccles JC. The self and its brain: an argument for interactionism. Berlin: Springer; 1977.

Prigogine I. La fin des certitudes Temps, chaos et les lois de la nature. Paris: Odile Jacob; 1996.

Rescher N. Objectivity. The obligations of impersonal reason. Notre Dame, IN and London: University of Notre Dame Press; 1997.

Roth PA. Meaning and method in the social sciences: a case for methodological pluralism. Ithaca, NY: Cornell University Press; 1987.

Salton G, McGill MJ. Introduction to modern information retrieval. New York: McGraw-Hill; 1983.

Schutz A. The social world and the theory of social action. In: Braybrooke D, editor. Philosophical problems of the social sciences. New York: Macmillan; 1965.

Schmuckler MA. What is ecological validity? A dimensional analysis. Infancy. 2001;2:419–36.

Seely BE. The scientific mystique in engineering: highway research in the bureau of public roads, 1918–1940. Technol Culture. 1984;28(4):799–803.

Shapley H. Of stars and men: human response to an expanding universe. Westport CT: Greenwood Press; 1964.

Smith VL. Microeconomic systems as an experimental science. Am Econ Rev. 1982;72(5):923–55.

Strevens M. Scientific explanation. In: Borchert DM, editor. Encyclopedia of philosophy, vol. 2. Detroit: Macmillan Reference; 2006.

Szabó ZG. Compositionality. In: Zalta EN, editor. The stanford encyclopedia of philosophy; 2013.

Tepe P. Kognitive Hermeneutik. Würzburg: Köningshausen & Neumann; 2007.

Vamanu I. Hermeneutics: a sketch of a metatheoretical framework for library and information science research. Info Res. 2013;18(3):S08.

Vattimo G. La società trasparente. Milano: Garzanti; 1989.

Weber M. Subjectivity and determinism. In: Giddens A, editor. Positivism and sociology. London: Heinemann; 1974.

Wilde L. On the use of laboratory experiments in economics. In: Pitt J, editor. The philosophy of economics. Dordrecht: Reidel; 1981.

Wilk RR. The impossibility and necessity of re-inquiry: finding middle ground in social science. J Cons Res. 2001;28(2):308–12.

Wittgenstein L. Philosophical investigations. In: Anscombe GEM, Rhees R, editor. Anscombe GEM (trans.). Oxford: Blackwell; 1953.

On the Development of Engineering Assets—The MATRIX Scheme

Frank A. Coutelieris and Antonios Kanavouras

This chapter presents a developmental approach through a well-defined 12-fold classification matrix scheme, recognized via a synthetical mathematical analysis. Starting from the description and delimitation of the system and following a solid work flow scheme, we conclude on developed systemic classes, which at least, do have satisfactory empirical connections, mathematical sense and engineering significance. Towards this aim, the problem of similarity among physical phenomena is also discussed, not only in terms of philosophy but also by mathematical treatment. For those readers that are not familiar with the similarity concept, a relative specific section has been added just after the introductive one. Precisely, all the existent perceptions of a physical phenomenon constitute a four-dimensional vector space, on which a specific non-linear mapping might be applied to strictly classify the existing knowledge about the phenomenon in question. The latter becomes able through a classification matrix defined by the categorical descriptors (lines) and the levels of these descriptors (columns). This mathematical approach actually develops an engineering tool, being applicable on general problems of engineering interest. Later in this chapter, some applications of this scheme are presented, to underline the power of this concept as multi-purpose engineering tool.

© The Author(s) 2018   81
F. A. Coutelieris and A. Kanavouras, *Experimentation Methodology
for Engineers*, SpringerBriefs in Continuum Mechanics,
https://doi.org/10.1007/978-3-319-72191-0_4

The structure of this Chapter is as follows

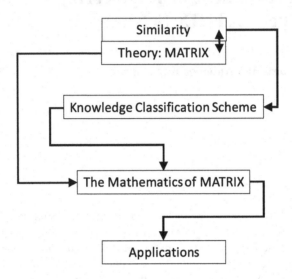

Introduction

Several scientific fields have commonly applied empirical and transcendental research approaches, mainly focusing on describing the identity and/or the differences of the object under research. The research outcome has been accordingly derived and used to confirm the validity of a scientific hypothesis via the logical combination of empirical observations, and conclusive theoretical and practical . However, commonly selected justification means are mainly picturing incidental events as observed within certain time intervals. Objectivity comes in degrees. Claims, methods and results can be more or less objective, and, other things being equal, the more objective the better. The term "objective" often conveys a special theoretical force with it when it is used to describe something. The admiration of science among the public and the authority science enjoys in public life stems to a large extent from the view that science is objective or, at least, more objective than other modes of inquiry. Understanding scientific objectivity is therefore critical to understand the nature of science and the role it plays in society.

In an attempt for enhancing the empirical observations and/or their correlations, significant data accumulation usually occurs, which may be quite well matched with, and depended on, the pre-defined justification means. Clearly, measurement does not result in a "view from nowhere", nor are typical measurement results free from presuppositions. Measurement instruments interact with the environment, and so results will always be a product of both the properties of the environment we aim to measure as well as the properties of the instrument. Thus, instruments provide a perspective view on the world (Giere 2006). Furthermore, incommensurability is rather inevitable for meanings, criteria and wit and could impact scientists' commitment against their scientific field. The semantic richness of scientific objectivity is also reflected in the multitude of categorizations and subdivisions of the concept (Megill 1994; Kuhn 1962; Feyerabend 1962).

Undoubtedly, the continuously increased numbers of observations obtained with advanced technological equipment adopting modern techniques, is used to support the possibility for something to be recorded as such. On the other hand, since the confirmation possibility level is the ratio of samples taken over all potential samples ($P_c = P_s/P_{st}$), indicates that the confirmation possibility P_c tends to 0 since the denominator tends to infinite (Popper 1963; Thornton 2009). Consequently, investigating the possibility of something not to take place, rather than the possibility of something to happen, seems more logical.

Additional puzzles arise when attempting to induct the empirical justification into a theoretical interpretation. Measurement and quantification help to reduce the influence of personal biases and idiosyncrasies and they reduce the need to trust the scientist but often at a cost. Standardizing scientific procedures becomes difficult when their subject matters are not homogeneous, and few domains outside fundamental physics exist. Although vastly applied, statistical analysis results could still be quite contradictory to the potential meaning and significance of the independent phenomenological events and individual steps or causes. A smooth connection of the individual data points, although provides the "general" evolutionary trend, still lets any data differences potentially "diluted" in the general "trend" of the curves, or the statistical significance level applied on the recorded justification differences. Conclusions are consequently drawn from such a mathematical and statistical analysis, while further geometrical treatments on the curves—providing additional estimations, for instance in the case of the kinetic parameters and phenomenological rates—are applied. In this sense, the treated data correlates its outcome with the justification targets, making the conclusions highly biased to the selected formulas a scientist is using in treating the data.

Usually, advanced mathematical models have also been developed to facilitate the predictive behavior of the experimental parameters in question. Direct models are simple and empirical, commonly deriving via a quantitative description of each phenomenon involved. They are empirical in nature, a characteristic which may safely allow the presumption that their parameters have no particular physical meaning; they are only used to interpolate data by means of a curve fitting procedure. On the other hand, mechanistic models are difficult both to derive and to handle, but most importantly, they are generally predictive and can thus be used for product/research design purposes. Evidently, modeling can be a powerful tool and significant effort has been put on the development of such a methodology.

Additional topics to consider in a scientific study, such as complexity of multifactorial systems, possess a number of intrinsically coherent properties. Studying them on an abstract but also in-depth, at erudite and meticulous applications, requires before anything else to establish the proper experimental conditions. In a more general statement, for justifying the hypothetical potential evolution of a biologically active system, a procedural rather than a "static" method through and around the system has to be established. Consequently, the experimental conditions can only be the defined and applied particular courses of actions intended to achieve a result.

Accordingly, it is important to study the cohesions among the chronologically disordered properties of a system, in order to validate, predict and eventually manage the resistance of the system against disclaiming of the hypothesis. By focusing on the system's principal properties, it would be possible, in due course, to approach a knowledge based evidence, proving the lowest letdown risk of the hypothesis, in generalized universal terms, avoiding a rather incidental image of the momently inertia of the system's complexity. Issues related to the nature of determinism, such as how we understand that something is a cause of any phenomenon, the limits of determinism and the structure of scientific explanation, is what constitutes a scientific answer when we put the question "why" instead of "what" (Weber 1949; Wittgenstein 1953).

Similarity

Scientists may investigate a phenomenon in order to support a theory, since theories may have gaps and inconsistencies that need to be completed. The major tool for obtaining experience in the physical world is based on reproducing the physical phenomena in a controlled environment, i.e. within a lab or filed experimentation procedure, is actually based on similarities between phenomena or classes of physical phenomena, whose identification lies among the fundamental considerations of a physical scientist in charge (Kroes 1989). This reproduction should be in a standard and repeatable manner (Hacking 1983). In fact, the perception of a phenomenon is deriving in the context of a hypothesis via the logical combination of empirical observations, and conclusive theoretical and practical interpretations (Kuhn 1962). Therefore, there is an infinite number of phenomena, which claim to represent an objective reality with a higher or a lower precision level.

Therefore, the ability to correctly infer the values of quantities in one physical system from knowledge of the values of the quantities in a system that is physically similar to it, rests on the ability to correctly establish that the two systems are similar. Towards this aim, the necessity of an existent classified knowledge has already been accepted (Sterrett 2002, 2006).

A powerful tool assisting this classification is the similarity, that assures and requires a mapping between the various phenomena. Once two systems have been identified as similar, one may be able to draw a similarity mapping to keep certain relationships about quantities in one system the same as in the other (Sterrett 2006). Precisely, the concept of similarity assures and prerequisites that a mapping between elements of the system exists, having also the property to preserve the structure of the situation that is relevant to the phenomenon of interest. In other words, when two systems/states are similar, a similarity mapping must be able to be drawn based on keeping certain relationships, quantities or elements in one system the same as in the other.

Observations of events expressed in the laboratory are considered informative about things and events that go beyond the specifics of the observed case, in support of a broader theoretical approach. This is possible due to the assumption that there is a class of events or situations that are similar to a given event, and that a given event is informative of other events in that class. A question now arises (Glymour 1970): what determines the class of other events to which an observation is made on a specific setup, is deemed similar? The answer apparently, refers to a comparison of various considerations of the same object or to a comparison among various objects on a common hypothesis (Sterrett 2006).

Specific well posed criteria are necessary to identify similarity. For that it is widely accepted that a classification of the existing knowledge is necessary (Sterrett 2006). However, a major difficulty to that is the definition of the specific similarity criteria, in order to assure that similarity indeed exists. Evidently, a researcher applying improper similarity criteria might obtain erroneous or meaningless results, yet, still spending significant amounts of resources. Accordingly, poor similarity criteria could lead to a research duplication (and, therefore, to an effort's and resources' waste), due to the lack of a deeper insight on the existing experience/ knowledge regarding the phenomenon under consideration.

Theory

Given a physical system, creating and applying certain technical restrictions at various levels during the experimental phase, could control the expression of the systemic properties, along with the processes' outcome. The ably combination of the outcome data obtained under controlled conditions shall provide a holistic approach of the system, which will reveal the cohesive points of the phenomena as occurring in a well-defined system. Such a methodology requires sensitive, accurate and inexpensive "indicators" to provide the most accurate shelf life prediction, by considering (monitoring of the product) those major parameters influencing the deteriorative reactions.

Primarily, we should define and indicate the functional ways of organizing the system under investigation and impending its inner logic into the preservation hypothesis study attempt. Therefore, we shall call "judgment" this inner logic approach. More specifically, judgment is here instated as either "analytical", the one which exposes the systemic phenomena in a sense that their opposite sites lead to logical contradictions, or "combinatory", the judgment which ultimately reinforces the hypothesis. Furthermore, we shall call "proposals" the valid ways for impending this logic. The a posteriori proposals are those provide the contingency, while the a priori proposals are those based on the necessity and universality, which, therefore, makes them non-revisable. Overall, as it is not the need for a judgment that determines its content, each proposal may be either analytical or synthetical.

A combination of the above in a mathematical mode, establishes four definitions, consistently resulting in the following functions of judgment:

(i) a posteriori-*analytical* —(verification model),
(ii) a priori-*analytical*—(predictive model),
(iii) a posterior-*combinatory*—(validation model),
(iv) a priori-*combinatory*—conditions for the experience possibilities, engineering tool).

Furthermore, since the current work overall research approach is focusing on setting the potential conditions for a functional engineering tool, we are in a need of a supportive methodological approach, briefly summarized in the following Table 1.

Table 1 presents the sequentially established functions of judgment towards developing the a priori-combinatory final judgment. All the work flow, namely the epistemic structure, is also presented. In specific, the table portrays the developmental roadmap, divided in three (3) stages corresponding to equally numbered deliverables. At each development Stage, the universal scientific philosophies for criticizing the system (systemic principles), along with the means and techniques to practically structure the systemic cohesions (proctorship), shall implicitly support the judgment functions which will structure the critically approached milestones (deliverables) and lead to the ultimate model.

Due to the fact that knowledge cannot precede the experience, the first step from Stage 1 to Stage 2, has to consider both the physical principles of the system along with an empirical proctorship. That is actually an a posteriori-*combinatorial* step to obtain a physical description of the system. Needless to say, the set-up of the experimental system and the forms of surveillance *in-space* and *in-time* must secure such a mode. From a certain point of view, it would be the model's rationality that shall assist an "economic" experimentation.

The next step brings about the cerebration level for recognizing and classifying the inherent systemic "ways-of-order", captured in a categorical description of the system, simply designated as the "categories".

The a posteriori-combinatorial analysis and the categorical descriptions together constitute the perception of the system and through that necessitate an inherent mathematical relativistic expression of the systemic cohesions existing therein. The mathematical expression evidently incorporated the principles and the logic, and consequently it could be trusted as a non-revisable reinforcement tool. Once this is completed, the food-packaging system may require a certain re-construction, via

Table 1 The epistemic structure of the work organization scheme

Stages	1	——>	2	——>	3
Principles		Physical	Intellectual	Mathematical	Engineering
Proctorship		Empirical	Logical	Morphic	Relativistic
Functions of judgment		a posteriori-combinatory	Reinforcement	Necessity	a priori-combinatory
Deliverables	System		Categories		Model

engineering of the systemic properties, in the direction of reducing the risk of the preservation disclaim.

At Stage 1 we have the detailed description of the system and the definition of the phenomena occurring there. Furthermore, the parameters affecting the results must also be identified and stated at this Stage. The Deliverable in Stage 2 is the Categories, the main descriptive systemic components, which form the hypothesis classification scheme. Its necessity and success are respectively originated from the need and aims capably applied to correctly describe the system. The description will be according to the eventual reflections in which intellect composes the phenomena. Thus, this Stage is solely a formal act of human cognition. Finally, Stage 3 is the use of the above two stages for a specific engineering problem.

Knowledge Classification Scheme

The necessity and success of a classification scheme is originated from the need, and aims at the applied capability to correctly describe the system, based on the eventual reflections of the way our mind composes the phenomena. The overall classification scheme has to cover the hypothesis' context possibilities for the system in question and regulate the system boundaries (classification frame).

Currently, knowledge classification attempts have been, in most of the cases, rather empirical, without explicitly defined rules, seemingly following a time-dependent evolution of inexplicable form (Feyerabend 1962). As far as the existence of any phenomenon cannot be known a priori but only through determinate cognition of experience, it has not been able to identify in what respect the empirical intuition of it, would be distinguishable from that of others. Furthermore, a general principle of all analogies rests on the synthetical unity of all phenomena according to their relation in time (Kant 2014). At this point, it is mathematics, according to the rules of a mathematical synthesis that could allow a representation of the perception of a phenomenon. In this context, mathematical judgments are always synthetical, while it was found that mathematical conclusions proceed according to the principle of contradiction, which the nature of every proven certainty requires (Troelstra and Schwichtenberg 2000).

It is essential to primarily consider the conceptual description of the system, still in the light of adequacy for a food-packaging system model establishment. The basic "in-process-out" context was converted into the necessary justification methodological approach by incorporating the optimum and minimal, yet straightforward systemic participants. In simple terms, the basic context turned into the following principally described oxidation hypothesis, i.e. the categorical descriptors:

$$\text{matter} + \text{energy} \xrightarrow{\text{relationships}} > \text{outcome} \tag{1}$$

All categories have to satisfy the hypothesis' context possibilities for the system in question. Having adequately considered and investigated all levels and categories, it is necessary to regulate the boundaries of the model via impacting on each and every systemic participant, *in-principal*. Conclusively, the classification will also reflect the theoretical solidness of the model, independent of the empiric experience, in which case the model shall be capable of assessing the shaped categorical imperative of the hypothesis.

Following the above, the categorical description of the system was developed. We concluded an adequate classification frame of—at least—a 12 cells scheme, consisted of 4 main structural categories extended up to 3 empirically defined levels. These levels denote the degrees of freedom for a set of conditions within a critical judgment. In practice, all levels of a given category condense the impact of this systemic participant where it is the cohesion among the systemic orders that are also considered. It is worth-mentioning that the amount of the cells is very well defined in terms of mathematics: the rows are 4 because of the 4 quantities involved in Eq. (1) while the columns are 3 because 3 is the least acceptable number of linearly-independent points that could follow a non-linear relationship. Overall, it is all four categories at all three levels that should contain all universal hypothesis' possibilities for a food system during its shelf-life. For the defined food-packaging system, the four specific categories and their corresponding general levels are given in the following Table 2.

Each individual "category-level" cell represents the particular conditions for a systemic activity. In a sense, each cell is a "hypothesis disclaiming class", a structural contributor to the possibility of the overall hypothesis disclaim. Furthermore, summary of the classes embraces the cohesion within the system context, rather than the independent interpretational risks due to the theoretical impregnation of the observations. Cross sections of the classification cells actually provide the much-needed space for inter-subjectivity, or i.e. they allow for the maximum objectivity in validating the hypothesis. The common ground among all classes' interactions shall eventually deliver the basic principles behind the phenomena, the cohesions in the food system, the only physically meaningful and responsible, potential conditions of knowledge.

Each column of the above matrix within the levels of the categories, represents a specific situation within the system. The general trend is the increase of the description complexity, when moving from left to right. The first column refers to only one object/variable as a major representative for describing the system. Following a conservation law and/or a relative mass/energy balance, only one mathematical equation seems adequate to describe what takes place in the system.

Table 2 The general categories and their corresponding general levels classification frame

Categories	Levels		
Matter (Quantity)	One	Many	All
Energy (Quality)	Reality	Disallowance	Restrictions
Inherent potential (Relationships)	Inter-dependent	Reasons	Intra-dependent
Outcome (Processes)	Potential	Existence	Necessity

A single factor might be as well selected to describe the macroscopic behavior of the system, on the basis of one specific relationship between the variables selected and the outcome quantity produced. In brief, the first column of the matrix refers to *one* variable involved in *one* algebraic, differential or integral equation that produced by applying *one* fundamental principle in the system, while *one* quantity is selected to macroscopically describe the system. This column produces a rather primitive ideal outcome, which can roughly represent the system.

The mid-column is produced by the transition from the one-dimensional events to multi-dimensional ones, with finite dimension. This vector-space dimension might represent the number of variables selected to describe the system (matter) or the details on the phenomena occurred (energy/relationships) or both. In any case, a system of differential or algebraic equations is produced by applying the corresponding fundamental principles on the system—parameters and reactivity-, while a single *one* parameter is selected to macroscopically describe the system. Although a single macroscopic outcome is defined, the difference from the first column's outcome is significant, since this second quantity includes the inter-effects of more variables and parameters, being therefore more accurate in satisfying more efficiently the approach.

Finally, the third column describes the system in infinite dimensions. That signifies an infinite number of variables, which must be treated through asymptotic techniques, since it is not possible to define a system of equations with infinite size. The selection of *one* macroscopic quantity which is not only adequately describing the system's behavior (despite the problems arisen due to infinite dimensions), but also considers all the parameters impact (although not necessarily known in full details). The overall concept is described in the following Table 3.

Specifically, the first column in the above matrix represents a simple, one-dimensional description of the hypothesis, the second column corresponds to a next level transition in a multi-dimensional space while the third column depicts the influence and the cohesions in an infinite multi-dimensional vector space. Regarding the macroscopic description of the phenomenon/-a occurring as part of the systemic behavior under certain conditions, both first and second columns denote the selection of one representative quantity, but they clearly differ in the amount of parameters whose influence has to be taken in account: the quantity of the first column represents unavoidably the effect of one parameter, in the quantity

Table 3 Translation of Table 2 in the world of mathematics

Categories	Level 1	Level 2	Level 3
Dimension	One	Finite	Infinite
Mathematical treatment	Equation	System of equations	Asymptotic
Macroscopic quantity	One (produced by the solution of the equation)	One (produced by the solution of the system)	One (appropriately selected)

of the second column a finite number of parameters is assumed to affect, while the quantity of the third column incorporates an infinite amount of parameters as well as their impact.

Decisive Rules

It is important now to define the rules regarding the matrix fill-in. These rules shall eventually assure a safe approach to an acceptable model for engineering purpose. Through this procedure, it should be able to define the transition functionality and efficiency from the first to the third column, in practice and within the empirical reality available.

The main rules for filling the matrix are:

Rule 1: The transition from a description of a system to a model is able if and only if, all the cells of Table 2 are appropriately filled according to a given hypothesis.

Rule 2: A system may allow for more than one transition pathways from description to model, as the content of the cells in Table 2 are not obligatory unique.

Rule 3: If any cell in Table 2 contains multiple values, the selected macroscopic quantity has to be different, in accordance to the selected parameters. Although all of the potential different quantities in a cell are totally equivalent among each other, it is always possible to interchangeably translate each one of them to another through a simple relating process.

Subsequently, the completion of the twelve cells in Table 3, transforms the systemic mathematical description into an engineering tool that is self-confidently obtained through a well-established methodology of the available knowledge classification. Via this mathematical shelf assessment and development of the engineering model, the whole system and the classified knowledge on that, is screened for riffles and open points that may not allow for proper and adequate compliance of theory to mathematics, through the experience (field/lab observations). The above-mentioned process at the same time may reveal the experimental technology needed in order to answer and complete the inconsistencies in knowledge about the phenomenon under question.

The classification matrix has the ambition to convey the systemic physics, from the mathematical description, to engineering applications, though the systemic logical approach. The mathematical expression for bonding these classes together can convincingly asses the hypothesis disclaim as such a model would have been employed on the interfaces and common ground of the system principles. From a practical point of view, the matrix confirms the experimental design that provides the outmost pragmatic input for an engineering model.

On the base of the above, we worked towards identifying the mathematical expressions among the potential conditions, revealing the necessity of the system.

The resulted mathematical expressions will work on both confirming as well as assessing the adequacy of the knowledge of the system we have been collecting. We have considered the term "quality", as the level of consumers' satisfaction, the term "shelf life" as the possibility of disclaiming this satisfaction and the term "preservation" as the possibility of disclaiming the shelf life. Utilizing this approach, a shelf-life engineering model could result for the maximization of the preservation possible conditions, the minimization of the shelf life disclamation and maximum quality within this system.

The Mathematics of Matrix

Going back to Eq. (1), the four, to be called categorical descriptors (namely, matter, energy, relationships and outcome), express all of the phenomenon characteristics. To further elaborate on this, categories are considered as conceptions of an object, by means of which its intuition is contemplated as determined in relation to one of the logical functions of judgment. These descriptors are independent to each other since matter and energy are obviously axiomatic entities, the relationships exist independently on the object they are applied on, while the outcome, although seems to be related with matter and energy through the applied relationships, is independently selected through the hypothesis accompanying the investigation of the phenomenon. Therefore, outcome is mathematically independent on the other three elements. From the knowledge point of view, the above Eq. (1), expresses the currently available experience about the phenomenon. Hence, this application results to an infinite number of four-component vectors of the form $\{m, e, R, o\}$, each one describing a specific perception of the phenomenon. All these vectors actually define the set V, each component describing the precise *value* of the categorical descriptor used.

In support of the above, an example of engineering interest, could be the problem of instantaneous sorption of a substance "A" into a solid media. The "A" is assumed to be diluted in a Newtonian fluid flowing towards the solid surface under laminar flow conditions. When no reaction among the media is assumed, the available mass transport mechanisms are the convection (i.e. mass transport due to the motion of the medium) and diffusion (i.e. mass transport due to concentration gradients/differences), which have been mathematically described through the well-known convective diffusion equation (Bird et al. 1960). Recognizing the time-dependent spatial distribution of the concentration of "A" as the desirable outcome, results into a typical vector $\underline{v}$, (which is one of the above-mentioned vectors):

$$\underline{v} = \left\{ A, \ convective - diffusion, \ \frac{dC_A}{dt} + \underline{U}_A \cdot \nabla C_A = D_A \nabla^2 C_A \ with \ C_A(interface) = 0, \ C_A(\underline{r}, t) \right\}$$

$$(2)$$

At this point, it is important to clarify that every vector of V includes all the previously defined vectors, thus identifying the evolution of the knowledge about a specific phenomenon with the time. Therefore, a type of arrangement $\prec$ is defined through the time $\hat{t}$ when the perception $\underline{v}(\hat{t})$ has been formulated, as follows:

$$\hat{t}_1 < \hat{t}_2 \Leftrightarrow \underline{v}_1(\hat{t}_1) \prec \underline{v}_2(\hat{t}_2) \tag{3}$$

To further understand the arrangement $\prec$, it has to be mentioned that $\underline{v}_2$ in the above Eq. (3) contains all the knowledge existed in $\underline{v}_1$, since $\hat{t}_1 < \hat{t}_2$. In this context, every new perception of a phenomenon contains all the current knowledge about this phenomenon, plus a new contribution. Obviously, there are several cases where a newly obtained knowledge actually contradicts and eventually cancels a part or all of an existing knowledge on a phenomenon. In such a case, the existing knowledge is just proven as false knowledge. The overall achievement of proving the existing knowledge as actually false, can be treated as a new affirmative knowledge, by itself. In that sense, the arrangement previously defined in Eq. (3) is always valid, even for the case that a new knowledge negates any previous one.

Now, let's define the internal operation $\oplus$ as follows

$$\forall \underline{v}, \underline{w} \in V \, \exists \underline{u} \in V : \underline{u} = \underline{v} \oplus \underline{w} = \begin{cases} \underline{v} & \text{if } \underline{w} \prec \underline{v} \\ \underline{w} & \text{if } \underline{v} \prec \underline{w} \end{cases} \tag{4}$$

The above process identifies existent accumulated experience about a phenomenon under investigation, following any recent scientific contribution towards its knowledge. The specific relationship (operation) defined through Eq. (4) is commutative and associative, while it includes an identity element as well as inverse elements. For those readers interested in, detail proofs are again given in a separate specific section in the end of this chapter. Still, it is important to underline that the above-mentioned accumulation also includes fractions of knowledge that may, partially or totally, negate the existing knowledge. In terms of mathematics, this accumulation represents a series where each term is accompanied by its own particular sign.

By defining the amount of the accumulated knowledge included in the vector $\underline{v} \in V$, $\lambda = \|\underline{v}\| \in \mathbb{R}$, as the regular norm of the vector, is able to calculate the evolution ratio between every two elements of V. If λ_i and λ_j are the amounts of knowledge embedded in $\underline{v}_i \in V$ and $\underline{v}_j \in V$, respectively, then

$$\mu_{ij} = \frac{\lambda_i}{\lambda_j} = \frac{\|\underline{v}_i\|}{\|\underline{v}_j\|} \tag{5}$$

Obviously, $\mu_{ij} > 1$ when $\underline{v}_j \prec \underline{v}_i$, while $\mu_{ij} < 1$ when $\underline{v}_i \prec \underline{v}_j$.

Now, let's define the external operation $\times$ as follows:

$$\forall \underline{v}, \underline{w} \in V \, \exists \mu \in R : \underline{w} = \mu \times \underline{v} \Leftrightarrow \mu = \frac{\|\underline{w}\|}{\|\underline{v}\|} \tag{6}$$

The above operation actually quantifies the relative significance of the knowledge evolution through any two perceptions of a phenomenon under investigation. Operation $\times$ defined through Eq. (6) presents compatibility with scalar "multiplication", satisfies the distributivity of $+$ over $\times$ as well as distributivity of $\times$ over $\oplus$. For those readers interested in, detail proofs are again given in a separate specific section.

The above definitions and properties guarantee that the structure $\{V, \oplus, \times\}$ is a vector space of a basis containing the four vectors $e_m = \{m, 0, 0, 0\}$, $e_e = \{0, e, 0, 0\}$, $e_R = \{0, 0, R, 0\}$ and $e_o = \{0, 0, 0, 0\}$. Obviously, the dimension $= 4$. To prove that the above structure is indeed a vector space, it is necessary to show that (a) the elements e_m, e_e, e_R and e_o are linearly independent, and (b) that they might produce the whole vector space. Indeed, the categorical descriptors defined in Eq. (1) are independent to each other, because there is no straightforward transformation to produce anyone of them as a linear combination of the others three. For the matter, the energy and the relationships, this is rather obvious. On the other hand, the liberty of selecting any appropriate macroscopic quantity to represent "outcome" assures that this descriptor is independent on the others three. Finally, it is rather obvious that any vector of V is a linear combination of e_m, e_e, e_R and e_o.

It is now straightforward to define a mapping m_p^{ln} on this vector space, as follows

$$\mathbb{R} x V \rightarrow M_{3x1}(V) : m_p^{in}(\underline{v}) = \{ \lambda_1 x \underline{v}, \quad \lambda_2 x \underline{v}, \quad \lambda_3 x \underline{v} \} \tag{7a}$$

with

$$\lambda_1 \rightarrow 0 \tag{7b}$$

$$\forall \lambda_2 \in \mathbb{R} \, \exists M > 0 : \lambda_2 < M \tag{7c}$$

$$\lambda_3 \rightarrow +\infty \tag{7d}$$

The first element of the mapping, $\lambda_1 \times \underline{v}$, represents a nearly zero amount of knowledge, the second, $\lambda_2 \times \underline{v}$, represents any finite amount of knowledge and the third one, $\lambda_1 \times \underline{v}$, the almost total infinite amount of knowledge that can me accumulated for the physical phenomenon under research. This classification is consistent with the philosophical wit of "one-many-all", encountered in the modern philosophy (Popper 1963).

The above mapping [Eq. (7a, b, c, d)] produces a matrix with four lines, each one standing for each of the elements $\{m, e, R, o\}$, and three columns, the first for the vector $\lambda_1 \times \underline{v}$, the second for the $\lambda_2 \times \underline{v}$ and the third for the $\lambda_3 \times \underline{v}$. In terms of rationalism, each column of this matrix represents a specific perception of the phenomenon, as previously presented in detail. Precisely, the first column refers to a vector containing the minimum non-zero knowledge of the phenomenon, where only one variable, along with only one mathematical equation produced by one simple conservation law or a relative mass/energy balance, are considered adequate to describe the specific perception. In fact, the first column of the matrix produces a rather primitive ideal outcome, which can roughly represent the phenomenon. The second column refers to the maximum finite knowledge currently available, where a finite number of variables are selected to describe the phenomenon and, therefore, a system of equations is produced, while a single one parameter is again selected to macroscopically describe the phenomenon. Briefly speaking, the second vector is a more accurate and more efficient representation of the phenomenon under consideration. Finally, the third column describes the absolutely holistic perception of the phenomenon, taking into account an infinite number of variables that define a system of equations with infinite dimension. In other words, the third vector describes the overall currently available knowledge about a phenomenon, identifying all the parameters' impact, although not necessarily known in full details.

It is important to note that

$$\lambda_1 x\underline{y} \prec \lambda_2 x\underline{y} \prec \lambda_3 x\underline{y} \tag{8}$$

i.e. the last column of the matrix includes all the knowledge embedded in the previous two columns. Although sounds valid at a glance, the direct use of only this third column is impossible without the use of the previous two, due to the high complexity of the description and the infinite quantities involved. In this context, the values of the mapping m_p^{ln} produce the necessary classification of knowledge through Eq. (7a, b, c, d). Apparently, the above mapping builds an internal similarity between the columns of matrix, as far as they are produced through the same mapping expression.

The aforementioned theory has been developed in order produce a tool for the detection of internal and external similarities under specific similarity criteria. For the application of such a theory, the development of a detailed methodology is crucial. In order to achieve potential internal similarity, it is necessary to complete the 4×3 matrix, i.e. to define a mapping of the form given by Eq. (7a, b, c, d). Obviously, there are more than one options (definitions) of such a mapping, therefore the matrix is not unique. What is important here is to carefully follow the decisive rules, previously presented in detail. Moreover, the use of this matrix allows for the identification of lacks of knowledge about the phenomenon under investigation: this lack exists if it is not able to fill all the cells of the matrix, i.e. whether is not able to define the three real numbers λ_1, λ_2 and λ_3 in Eq. (7a, b, c, d).

Applications of the MATRIX Scheme on Engineering Problems

This chapter presents two engineering applications of the MATRIX scheme, as discussed previously. Precisely, the first engineering problem addressed by the application is the auto- and photo- oxidation of olive oil, when packaged in several different packaging materials and stored under several conditions. The second application refers to the mass transport through granular porous materials, under several adsorption mechanisms.

Application I: Oxidation of Packaged Olive Oil

Consider a certain mass of extra virgin olive oil, packaged in glass or plastic bottles, and stored under quite a range of practically potential temperatures (5–40 °C), in presence or in the absence of light. After a short quiescent period, photo-oxidation reaction takes place when light is available, while auto-oxidation due to oxygen entering the bottle also occurs, mainly in oxygen permeable containers such as plastics. These oxidation phenomena significantly lower the quality of the oil, which reaches the end of its shelf-life after a definite time period. According to the above brief description, Table 2 can be completed as follows:

A thorough description and relevant literature references in support of the selections are given below (Table 4).

- Line 1—Matter: For the single dimensional case (Column 1), we shall consider oxygen concentration, as the necessary species involved in the evolution of oxidation, (Korycka-Dahl and Richardson 1978). The oxygen concentration in the oil is dependent on the oxygen partial pressure in the headspace of the oil (Andersson 1998) affecting the amount of oxygen that is dissolved in the oil, with oil oxidation to increase with the amount of dissolved oxygen (Min and Wen 1983). Oil oxidation rate is independent of sufficiently high oxygen concentrations (Kacyn et al. 1983; Labuza 1971). However, the reverse was true at low oxygen pressure in the headspace (Karel 1992), where the oxidation rate is becoming independent of the lipid concentration, while temperature further promoted the oxidation rate as did also the presence of light and metals such as iron or copper (Andersson 1998). Researchers have also calculated the amounts of oxygen sufficient to promote the oxidation in oils as reported by Przybylski and Eskin (1988) and Min and Wen (1983), at various temperature ranges, suggesting the oxygen presence to oxidation by-products relationship. For example, for rapeseed oil in the dark, temperature and oxygen were highly correlated to the presence of 2-pentenal and 1-pentene-3-one relatively elevated temperatures, but not at 35 °C (Andersson and Lingnert 1999). Convection and solubility of the oxygen in the oil are other important pathways for oxygen penetration into the oil, affecting the oil oxidation at high temperatures due to low solubility of oxygen in the oil (Andersson 1998). By adding the significant species (fatty acids and the relative produced hyperoxides) as well as by taking into account the packaging material (glass, PVC, PET), it is able to move to a multi-dimensional space of finite dimension (Column 2), which is assumed to produce more accurate results. Commonly, extra-virgin olive oil is usually packaged in glass, tin or plastic bottles. The primary advantages for the two first

Table 4 Classification matrix for olive-oil oxidation

Categories	Level 1	Level 2	Level 3
Matter	Oxygen	Unsaturated fatty acids, hyperoxides, O_2, container material (glass/PET/PVC)	Hydroperoxides and deriving flavor profile compounds for T °C and O_2 concentration
Energy	Chemical reactions: auto and photo-oxidation for deriving hyperoxides are as follows: $O_2 \xrightarrow[hv]{I} |k_a O_2^\circ$ $RH + O_2^\circ \xrightarrow{k_b} ROOH$ $RH + O_2 \xrightarrow{k_c} ROOH$	Chemical reactions of autoxidation: $R_2\!-\!CH\!=\!CH\!-\!\underset{\underset{OOH}{\mid}}{CH}\!-\!R_1$ $R_2\!-\!CH\!=\!CH\!-\!\overset{B}{\underset{\underset{O\cdot}{\mid}}{CH}}\!-\!\underset{\mid}{CH}\!-\!R_1$ $R_2\!-\!CH\!=\!CH \cdot + \ OHC\!-\!R_1$ And photo-oxidation: $O_2^\circ + O_2^\circ + 2H^+ \rightarrow H_2O_2 + O_2$ $H_2O_2 + O_2^\circ \rightarrow HO^\circ + OH^- + {}^1O_2$	Autoxidation: $R_2\!-\!CH\!=\!CH \cdot + \ OHC\!-\!R_1$ $R_2\!-\!CH\!=\!CH\!-\!OH$ $R^2\!-\!CH^2\!-\!CHO \quad R^2\!-\!CH\!=\!CH^5$ Photo-oxidation: ${}^1O_2 + ROOH \rightarrow$ allyl hydroperoxides
		Fick's Law for packaging materials oxygen diffusion: $j_{O_2} = -D_{O_2,bottle} \nabla C_{O_2}$	For each species i, Fick's Law for packaging materials oxygen and hyperoxides diffusion: $j_i = -D_{i,bottle} \nabla C_i$
		Mass flux of oxygen in packaging: $j_{O_2} = U_{O_2} C_{O_2}$, and its velocity (Darcy's Law): $U_{O_2} = -K_{O_2} \nabla P_{O_2}/\mu_{O_2}$	Oxygen and hyperoxides mass fluxes for species i: $j_i = U_i C_i$ and their velocity (Darcy's Law): $U_i = -K_i \nabla P_i/\mu_i$

(continued)

Table 4 (continued)

Categories	Level 1	Level 2	Level 3
Relation ships	Single reaction term mass transport equation: $\frac{dC_{hexanal}(t)}{dt} = k_a(T)C_{O_2}(t) + k_c(T)C_{what?}(t)C_{O_2}(t)$ with the initial and boundary conditions $C_{O_2}(t=0) = C_0$ $C_{ROOH}(t=0) = C_1$ $C_{hexanal}(t=0) = 0$	System of mass transport equations (diffusive, convective and reaction terms), where $i = 1, 2, \ldots, N$: $\frac{dC_i}{dt} + \underline{U}_i \cdot \nabla C_i = D_{i,mix}\nabla^2 C_i \pm \sum_{\substack{j=1 \\ j \neq i}}^{N} k_{i,j}(T)C_i C_j$ with the initial and boundary conditions	Oxidation species mass transport statistics via Life-cycle analysis in oil phase and from out-environment: $\frac{dC_i}{dt} + \underline{U}_i \cdot \nabla C_i = D_{i,mix}\nabla^2 C_i \pm \sum_{\substack{j=1 \\ j \neq i}}^{\infty} k_{i,j}(T)C_i C_j$ where $i = 1, 2, \ldots, \infty$ with the initial and boundary conditions
Outcome	Hexanal concentration $C_{hexanal}(\underline{x}, t)$	Hexanal concentration, averaged over space, as follows $\langle C_{hexanal}(t)\rangle = \int_V C_{hexanal}(\underline{r}, t)dr$	Possibility of oil not to reach the end of shelf-life (P_{safe}): $P_{safe}(t) = 1 - \dfrac{\int_0^{t_{critical}} \langle C_{hexanal}\rangle(t)dt}{\int_0^t \langle C_{hexanal}\rangle(t)dt}$

ones are its impermeability to gases, but the glass and plastic bottles had some disadvantages such as to favor the photo-oxidation (Méndez et al. 2007). Finally, to approach a vector space of infinite dimension (Column 3), it is necessary to consider all the species, i.e. to identify the factors of the system as a whole.

- Line 2—Energy: Following what previously stated for Column 1, the consideration of one important species (namely, hexanal) imposes a simple set of related chemical reactions (oxidation and photo-oxidations). On top of reactions, in the case of Column 2, where diffusion and convection must be also considered. The same also stands for the case of Column 3, for all the species that might be involved. Autoxidation of oils and the decomposition of hydroperoxides increase as the temperature increases (Shahidi and Spurvey 1996; St. Angelo 1996; Aidos et al. 2002; Marquez-Ruiz et al. 1996). A higher decomposition of primary oxidation products and formation of dimers or oxidized compounds, in addition to the production of hydroperoxides has been reported for higher temperature (Velasco and Dobarganes 2002). Temperature has also an effect on the oxidation of unsaturated fatty acids since they are more susceptible to the oxidation than saturated ones, primarily due to their low activation energy for the formation of fatty acid radicals (Przybylski et al. 1993), while antioxidants also decrease the activation energy of oil oxidation (Choe and Min 2006). Light accelerates the oil oxidation, especially in the presence of sensitizers such as chlorophylls, mainly by producing singlet oxygen (Lee and Min 1988), that it can be considered as much more important than temperature in $^1O_2$ oxidation, especially that of shorter wavelengths, (Sattar et al. 1976) and lower temperatures, (Velasco and Dobarganes 2002). On the other hand, temperature has little effect on $^1O_2$ oxidation due to the low activation energy requirements (Rahmani and Saari Csallani 1998).

- Line 3—Relationships: When one species is assumed to be produced only through chemical reactions (Column 1), only one governing transport equation, produced through a single mass balance, is used. In Column 2, an amount of N species is considered, therefore N equations are necessary to mathematically describe the system. Following the previous analysis for Line 2, these equations should moreover include both convection and diffusion terms, therefore form a system of N differential equations with N unknown species concentrations. Obviously, the above mathematical system has to be integrated along with the appropriate set of N initial and 2 N boundary conditions in order to assure the uniqueness of the solution. Briefly speaking, the transition from Column 1 to Column 2 takes place here by describing a system of equations rather than a single equation. Synergies and antagonistic effects inside the oil mass define the outcome of the system of edible oil which often contains multi-components interacting among them, thus altering the antioxidant performance when present together than when used separately (Brimberg and Kamal-Eldin 2003). The improved antioxidation of oil by a combination of components can also differ at the different oxidation stages (Choe and Min 2006). Thus, for the case of Column 3, we consider an infinite number of species involved at the same phenomena; therefore, an infinite system of equations can be produced. To overcome the barrier that such a mathematical system has no analytical solution or numerical approximation, it is necessary to describe it system through stochastically produced formulations, such as limit processes.

- Line 4—Outcome: In any case, the output must be a macroscopic indicator that adequately describes the quality of olive oil. Column 1 enforces the use of hexanal concentration for such an indication, since it is the only possible outcome. The selection of hexanal is based on studies regarding the decomposition products of hydroperoxides which have been considered responsible for the off-flavor in the oxidized edible oil also due to their threshold values affecting the consumer perception in a different impact compared to concentration of the compounds per se (Frankel 1985). Among those studies, hexanal, and 2-decenal and 2-heptenaland trans-2-octenal, were the major volatile compounds detected (Steenson et al. 2002). Kanavouras et al. (2004a), also separated and identified flavor compounds from differently preserved olive oil samples exposed mainly to fluorescent light and elevated temperatures that stimulated the oxidative alterations in the extra virgin olive oil packaged in various storage conditions (glass/PET/PVC bottles) for one year. Based on the abundance and evolution of individual flavor compounds, it was suggested that hexanal, 2-pentyl furan, (E)-2-heptenal, nonanal, and (E)-2-decenal were the compounds that most clearly described the oxidation. The transition to Column 2 implies spatial averaging in order to eliminate the dependence of hexanal concentration on the position in space. This quantity allows for calculations regarding shelf-life of the product. A predictive mathematical model was introduced to describe the mass transport from and to the oil phase through various packaging materials for several temperatures and light availability storage conditions (Coutelieris and Kanavouras 2006). By using the experimental results from a study of flavor compound found during storage of packaged olive oil, to develop a mathematical model (Kanavouras et al. 2004b) which showed that the evolution rate constants of hexanal clearly increased with temperature, while auto-oxidation reactions appeared to be less sensitive to changes in temperature. Finally, the incorporation of all the species (Column 3) allocates the definition and use of a more accurate quality index, namely P_{safe}. The latter expresses the probability for the olive oil to reach the end of its shelf life during a certain time period, being equivalent with the ratio of the areas below and above an arbitrarily defined quality threshold. This quality threshold depends on a certain acceptable value of hexanal concentration, considered as an upper limit for the quality acceptance. Therefore, P_{safe} is actually expressed by the ratio of the time-integrals of the spatially averaged hexanal concentration (outcome of Column 2).

Application II: Adsorption in Granular Media

Consider an assemblage of solid spherical grains that are able to adsorb a specific substance A. The void space between them is occupied by a neutral fluid, which contains the substance A. The solution is supposed to flow through void space (liquid phase), while the substance A approaches the solid interface due to convection and diffusion. Given a particular amount (mass) of A flowing through the granular porous medium, its portion that has been adsorbed must be estimated. Obviously, this portion depends on adsorption mechanism considered. The above brief description corresponds to a fully filled matrix as follows:

The highly significant point to underline for Table 5, is that it unfolds through levels. To further elaborate on that we shall consider a transition from left to right, in particular for each line:

- Line 1—Matter: For the single case (Column 1), we consider a substance A that is instantaneously adsorbed. The instantaneous adsorption is a very primitive but convenient assumption, allowing for a simple boundary condition on the solid-fluid interface (Levich 1962). By taking into account the solid surface (actually an adsorption mechanism interpreted into surface's properties), it is able to move to more detailed information about the system (Column 2) where detailed models referring to adsorption following isotherms have also to be adopted (Coutelieris et al. 2003). Finally, to attain a holistic approach (Column 3), it is necessary to describe adsorption as heterogeneous reaction, where the products (desorbents) behavior must be described, as well (Gavriil et al. 2014).
- Line 2—Energy: Although both columns include mass transport due to convection and diffusion, the transition through columns takes place with the different boundary conditions imposed as the adsorption mechanism becomes more and more complex.
- Line 3—Relationships: In any case, the convective diffusion equation is used to describe mass transport in the bulk face (Bird et al. 1960). As previously confirmed, the transition from Column 1 to Column 2 and 3 takes place due to the different considerations for the sorption mechanism. From the assumption of instantaneous adsorption (Column 1), where it is not necessary to describe the solid surface, we pass to Column 2, where it is assumed that adsorption takes place under a concrete isotherm. This implies the description of specific properties of the surface, playing the role of factors in the boundary condition produced by the expression of the isotherm. Finally, when the adsorption is described through heterogeneous reaction, as above stated for Column 3, a complicated condition including a lot of parameters, has to be used.
- Line 4—Outcome: The main output must macroscopically describe the adsorption. Column 1 enforces the use of concentration for such an indication, since it is the only possible outcome. The transition to Column 2 implies the dimensionless Sherwood number, which is an adequate index for the mass transferred between phases (Tien 1989). Finally, the most complex description (Column 3) uses the adsorption efficiency to illustrate the amount of A that has been sorbed by the solid grains (Coutelieris et al. 2004).

Table 5 Classification matrix for adsorption in granular media

Categories	Level 1	Level 2	Level 3
Matter	Substance A	Substance A, solid surface	Substance A, solid surface, products B_i produced due adsorption of A
Energy	Diffusion of A in the fluid phase. The fundamental principle is supposed to be Fick's Law, expressed as: $j_A = -D_A \nabla C_A$ Convection of A in the fluid phase. The mass flux is given as: $j_A = U_A C_A$ Instantaneous adsorption on the solid surface, expressed mathematically as $C_A(r=R) = 0$	Diffusion of A in the fluid phase. The fundamental principle is supposed to be Fick's Law, expressed as: $j_A = -D_A \nabla C_A$ Convection of A in the fluid phase. The mass flux is given as: $j_A = U_A C_A$ Adsorption following a specific isotherm for example, $D_A \underline{\mathbf{n}} \cdot \nabla C_A = \frac{k}{K} c_s$, where the parameters are defined by the Langmuir isotherm (fundamental principle): $\Theta_{eq} = \frac{K c_b}{1 + K c_b}$	Diffusion of all the species in the fluid phase. The fundamental principle is supposed to be Fick's Law, expressed as: $j_i = -D_i \nabla C_i$ Convection of all the species in the fluid phase. The mass fluxes are given as: $j_i = U_i C_i$ Heterogeneous reaction of first order $A \rightarrow B$, where B is assumed to desorbed in the bulk phase. The reaction rate is assumed to follow Arrhenius type, i.e. $R_n = k_0 e^{\frac{-E_A}{RT}} C_A$
Relation ships	Mass transport equation containing convective and diffusive terms, as follows: $\frac{dC_A}{dt} + \underline{U}_A \cdot \nabla C_A = D_A \nabla^2 C_A$ with the appropriate initial and boundary conditions	Mass transport equation containing convective and diffusive terms, as follows: $\frac{dC_A}{dt} + \underline{U}_A \cdot \nabla C_A = D_A \nabla^2 C_A$ with the appropriate initial and boundary conditions	System of mass transport equations, each containing convective and diffusive terms, as follows: $\frac{dC_i}{dt} + \underline{U}_i \cdot \nabla C_i = D_i \nabla^2 C_i$ with the appropriate initial and boundary conditions
Outcome	Concentration of A, $C_A(\underline{r},t)$	Overall Sherwood number $Sh_o = \frac{k_o L}{D}$	Adsorption efficiency $\lambda_0 = 1 - \frac{\iint_{S_{outlet}} c_A \underline{U} \cdot \underline{n} dS}{\iint_{S_{inlet}} c_A \underline{U} \cdot \underline{n} dS}$

Mathematical Proofs

For those readers that are interested in, this section presents the necessary mathematical proofs, regarding the vector space previously defined.

The operation $\oplus$ defined through Eq. (4) has the following properties:

- Commutativity $\forall \underline{v}, \underline{w} \in V \; \underline{v} \oplus \underline{w} = \underline{w} \oplus \underline{v}$

Proof: The result depends on the arrangement. If $\underline{w} \prec \underline{v}$ then $\underline{v} \oplus \underline{w} = \underline{v}$ and $\underline{v} \oplus \underline{w} = \underline{v}$, hence $\underline{v} \oplus \underline{w} = \underline{w} \oplus \underline{v}$. If $\underline{v} \prec \underline{w}$ then $\underline{v} \oplus \underline{w} = \underline{w}$ and $\underline{v} \oplus \underline{w} = \underline{w}$, hence $\underline{v} \oplus \underline{w} = \underline{w} \oplus \underline{v}$.

- Associativity $\forall \underline{v}, \underline{w}, \underline{u} \in V \; \underline{v} \oplus (\underline{w} \oplus \underline{u}) = (\underline{w} \oplus \underline{v}) \oplus \underline{u}$

Proof: The result again depends on the arrangement. Let's distinguish cases:

If $\underline{v} \prec \underline{w} \prec \underline{u}$ then $\underline{v} \oplus (\underline{w} \oplus \underline{u}) = \underline{v} \oplus \underline{u} = \underline{u}$ and $(\underline{w} \oplus \underline{v}) \oplus \underline{u} = \underline{w} \oplus \underline{u} = \underline{u}$, hence $\underline{v} \oplus (\underline{w} \oplus \underline{u}) = (\underline{w} \oplus \underline{v}) \oplus \underline{u}$.
If $\underline{v} \prec \underline{u} \prec \underline{w}$ then $\underline{v} \oplus (\underline{w} \oplus \underline{u}) = \underline{v} \oplus \underline{w} = \underline{w}$ and $(\underline{w} \oplus \underline{v}) \oplus \underline{u} = \underline{w} \oplus \underline{u} = \underline{w}$, hence $\underline{v} \oplus (\underline{w} \oplus \underline{u}) = (\underline{w} \oplus \underline{v}) \oplus \underline{u}$.
If $\underline{u} \prec \underline{v} \prec \underline{w}$ then $\underline{v} \oplus (\underline{w} \oplus \underline{u}) = \underline{v} \oplus \underline{w} = \underline{w}$ and $(\underline{w} \oplus \underline{v}) \oplus \underline{u} = \underline{w} \oplus \underline{u} = \underline{w}$, hence $\underline{v} \oplus (\underline{w} \oplus \underline{u}) = (\underline{w} \oplus \underline{v}) \oplus \underline{u}$.
If $\underline{w} \prec \underline{v} \prec \underline{u}$ then $\underline{v} \oplus (\underline{w} \oplus \underline{u}) = \underline{v} \oplus \underline{u} = \underline{u}$ and $(\underline{w} \oplus \underline{v}) \oplus \underline{u} = \underline{v} \oplus \underline{u} = \underline{u}$, hence $\underline{v} \oplus (\underline{w} \oplus \underline{u}) = (\underline{w} \oplus \underline{v}) \oplus \underline{u}$.
If $\underline{u} \prec \underline{v} \prec \underline{w}$ then $\underline{v} \oplus (\underline{w} \oplus \underline{u}) = \underline{v} \oplus \underline{w} = \underline{v}$ and $(\underline{w} \oplus \underline{v}) \oplus \underline{u} = \underline{v} \oplus \underline{u} = \underline{v}$, hence $\underline{v} \oplus (\underline{w} \oplus \underline{u}) = (\underline{w} \oplus \underline{v}) \oplus \underline{u}$.
If $\underline{w} \prec \underline{u} \prec \underline{v}$ then $\underline{v} \oplus (\underline{w} \oplus \underline{u}) = \underline{v} \oplus \underline{u} = \underline{v}$ and $(\underline{w} \oplus \underline{v}) \oplus \underline{u} = \underline{v} \oplus \underline{u} = \underline{v}$, hence $\underline{v} \oplus (\underline{w} \oplus \underline{u}) = (\underline{w} \oplus \underline{v}) \oplus \underline{u}$.

- Identity element $\forall \underline{v} \in V \; \exists \underline{0} \in V : \underline{v} \oplus \underline{0} = \underline{0} \oplus \underline{v} = \underline{v}$

Proof: Indeed, the vector $\underline{0}$ represents the nearly zero knowledge, which does not actually favor the evolution of the knowledge about the phenomenon.

- Inverse elements $\forall \underline{v} \in V \; \exists (-\underline{v}) \in V : \underline{v} \oplus (-\underline{v}) = (-\underline{v}) \oplus \underline{v} = \underline{0}$

Proof: Vector $(-\underline{v})$ in fact represents a singularity on the knowledge evolution, i.e. one point the knowledge of which tends to collapse all the existing knowledge. For instance, the Michelson-Morley experiment represents such a singularity regarding the 19th century knowledge about the atomic and sub-atomic physics.

The operation $\times$ defined through Eq. (6) has the following properties:

- Compatibility with scalar "multiplication" as follows: $\lambda x (\mu x \underline{v}) = (\lambda \mu) x \underline{v}, \forall \underline{v} \in V, \forall \lambda, \mu \in \mathbb{R}$

Proof: Let's $\underline{u} = \mu x \underline{v} \Leftrightarrow \mu = \frac{\|\underline{u}\|}{\|\underline{v}\|}$, therefore $\underline{w} = \lambda x(\mu x \underline{v}) = \lambda x \underline{u} \Leftrightarrow \lambda = \frac{\|\underline{w}\|}{\|\underline{u}\|}$. By combining both relations, we have $\lambda = \frac{\|\underline{w}\|}{\|\underline{u}\|} = \frac{\|\underline{w}\|}{\mu \|\underline{v}\|} \Leftrightarrow \lambda \mu = \frac{\|\underline{w}\|}{\|\underline{v}\|} \Leftrightarrow (\lambda \mu) \times \underline{v} = \underline{w}$, hence $\lambda \times (\mu \times \underline{v}) = (\lambda \mu) \times \underline{v}$.

- Distributivity of + over ×, as follows: $(\lambda + \mu) x \underline{u} = \lambda x \underline{u} + \mu x \underline{u}, \forall \underline{u} \in V, \forall \lambda, \mu \in \mathbb{R}$

Proof: Let's $\underline{u} = \lambda x \underline{v} \Leftrightarrow \lambda = \frac{\underline{u}}{\underline{v}}$ and $\underline{w} = \mu x \underline{v} \Leftrightarrow \mu = \frac{\|\underline{w}\|}{\|\underline{v}\|}$. By adding these two relations, we have $\lambda + \mu = \frac{\|\underline{u}\|}{\|\underline{v}\|} + \frac{\|\underline{w}\|}{\|\underline{v}\|} = \frac{\|\underline{u}\| + \|\underline{w}\|}{\|\underline{v}\|} \Leftrightarrow (\lambda + \mu) \times \underline{v} = \underline{u} \oplus \underline{w}$, hence $(\lambda + \mu) \times \underline{v} = \lambda \times \underline{v} \oplus \mu \times \underline{v}$.

- Distributivity of × over $\oplus$ as follows $\lambda x(\underline{u} \oplus \underline{w}) = \lambda x \underline{u} \oplus \lambda x \underline{w}, \forall \underline{u}, \underline{w} \in V, \forall \lambda \in \mathbb{R}$

Proof: The result depends on the arrangement as follows:

If $\underline{w} \prec \underline{v}$ then $\underline{v} \oplus \underline{w} = \underline{v}$ and $\lambda \times (\underline{v} \oplus \underline{w}) = \lambda \times \underline{v}$, while $\lambda \times \underline{v} \oplus \lambda \times \underline{w} = \lambda \times \underline{v}$, hence $\lambda \times (\underline{v} \oplus \underline{w}) = \lambda \times \underline{v} \oplus \lambda \times \underline{w}$.
If $\underline{v} \prec \underline{w}$ then $\underline{v} \oplus \underline{w} = \underline{w}$ and $\lambda \times (\underline{v} \oplus \underline{w}) = \lambda \times \underline{w}$, while $\lambda \times \underline{v} \oplus \lambda \times \underline{w} = \lambda \times \underline{w}$, hence $\lambda \times (\underline{v} \oplus \underline{w}) = \lambda \times \underline{v} \oplus \lambda \times \underline{w}$.

References

Aidos I, Lourenco S, van der Padt A, Luten JB, Boom RM. Stability of crude herring oil produced from fresh byproducts: influence of temperature during storage. J Food Sci. 2002;67:3314–20.

Andersson K. Influence of reduced oxygen concentrations on lipid oxidation in food during storage. Ph.D. thesis. Chalmers Reproservice, Sweden: Chalmers University of Technology and the Swedish Institute for Food and Biotechnology; 1998.

Andersson K, Lingnert H. Kinetic studies of oxygen dependence during initial lipid oxidation in rapeseed oil. J Food Sci. 1999;64:262–6.

Bird RB, Stewart WE, Lightfoot EN. Transport phenomena. New York: Wiley; 1960.

Brimberg UI, Kamal-Eldin A. On the kinetics of the autoxidation of fats: influence of pro-oxidants, antioxidants and synergists. Eur J Lipid Sci Techn. 2003;105:83–91.

Choe E, Min DB. Mechanisms and factors for edible oil oxidation. Compr Rev Food Sci Food Saf. 2006;5:169–86.

Coutelieris FA, Kainourgiakis ME, Stubos AK. Low Peclet mass transport in assemblages of spherical particles for two different adsorption mechanisms. J Colloid Interface Sci. 2003;264:20–9.

Coutelieris FA, Burganos VN, Payatakes AC. Model of adsorption-reaction-desorption in a swarm of spheroidal particles. AIChE J. 2004;50:779–85.

Coutelieris FA, Kanavouras A. Experimental and theoretical investigation of packaged olive oil: development of a quality indicator based on mathematical predictions. J Food Eng. 2006;73:85–92.

Feyerabend P. Explanation, reduction and empiricism, scientific explanation, space, and time. In: Feigl H, Maxwell G, editors. Minnesota studies in the philosophy of science, vol. III. University of Minneapolis Press; 1962.

Frankel EN. Chemistry of autoxidation: mechanism, products and flavor significance. In: Min DB, Smouse TH, editors. Flavor chemistry of fats and oils. Champaign, Ill: American Oil Chemists' Society; 1985. pp. 1–34.

Gavriil G, Vakouftsi E, Coutelieris FA. Mathematical simulation of mass transport in porous media: an innovative method to match geometrical and transport parameters for scale transition. Dry Techn. 2014;32:781–92.

Giere R. Scientific Perspectivism. Chicago, IL: University of Chicago Press; 2006.

Glymour C. On some patterns of reduction. Philos Sci. 1970;37:340–53.

Hacking I. Representing and intervening. Introductory topics in the philosophy of natural science. UK: Cambridge University Press; 1983.

Kacyn LJ, Saguy I, Karel M. Kinetics of oxidation of dehydrated food at low oxygen pressures. J Food Proc Preserv. 1983;7:161–78.

Karel M. Kinetics of lipid oxidation. In: Schwarzberg HG, Hartel RW, editors. Physical chemistry of foods. New York: Marcel Dekker; 1992. pp. 651–68.

Kanavouras A, Hernandez-Münoz P, Coutelieris FA. Shelf life predictions for packaged olive oil using flavor compounds as markers. Eur Food Res Techn. 2004a;219:190–198.

Kanavouras A, Hernandez-Munoz P, Coutelieris FA, Selke S. Oxidation derived flavor compounds as quality indicators for packaged olive oil. J Am Oil Chem Soc. 2004b;81:251–257.

Kant I. The critique of pure reason (Translated by Meiklejoh JMD). Adelaide: University of Adelaide Press; 2014.

Korycka-Dahl MB, Richardson T. Activated oxygen species and oxidation of food constituent. Crit Rev Food Sci Nutr. 1978;10:209–240.

Kroes P. Structural analogies between physical systems. Brit J Philos Sci. 1989;40:145–154.

Kuhn TS. The structure of scientific revolutions. Chicago: University of Chicago Press; 1962.

Labuza TP. Kinetics of lipid oxidation in foods. CRC Crit Rev Food Techn. 1971;2:355–405.

Lee EC, Min DB. Quenching mechanism of beta-carotene on the chlorophyll-sensitized photooxidation of soybean oil. J Food Sci. 1988;53:1894–95.

Levich VG. Physicochemical hydrodynamics. Englewood Cliffs, New Jersey: Pentice-Hall; 1962.

Marquez-Ruiz G, Martin-Polvillo M, Dobarganes MC. Quantitation of oxidized triglyceride monomers and dimers as a useful measurement for early and advanced stages of oxidation. Grasaa Aceites. 1996;47:48–53.

Megill A. Introduction: four senses of objectivity. In: Megill A, editor. Rethinking objectivity. Durham, NC: Duke University Press; 1994. pp. 1–20.

Méndez AI, Falqué E. Effect of storage time and container type on the quality of extra-virgin olive oil. Food Control. 2009;18:521–9.

Min DB, Wen J. Effects of dissolved free oxygen on the volatile compounds of oil. J Food Sci. 1983;48:1429–30.

Popper KR. Conjectures and refutations: the growth of scientific knowledge. New York: Harper 1963.

Przybylski T, Eskin NAM. A comparative study on the effectiveness of nitrogen or carbon dioxide flushing in preventing oxidation during the heating of oil. J Am Oil Chem Soc. 1988;65:629–33.

Przybylski R, Malcolmson LJ, Eskin NAM, Durance-Tod S, Mickle J, Carr RA. Stability of low linolenic acid canola oil to accelerated storage at 60 °C. Food Sci Technol (Lebens Wiss U Techn). 1993;26:205–9.

Rahmani M, Saari Csallany A. Role of minor constituents in the photooxidation of virgin olive oil. J Am Oil Chem Soc. 1998;75:837–43.

Sattar A, DeMan JM, Alexander JC. Effect of wavelength on light-induced quality deterioration of edible oils and fats. Can Inst Food Sci Techn J. 1976;9:108–113.

Shahidi F, Spurvey SA. Oxidative stability of fresh and heated-processed dark and light muscles of mackerel (Scomber scombrus). J Food Lipids. 1996;3:13–25.

St. Angelo AJ. Lipid oxidation in foods. Crit Rev Food Sci Nutr. 1996;36:175–224.

Steenson DFM, Lee JH, Min DB. Solid-phase microextraction of volatile soybean oil and corn oil compounds. J Food Sci. 2002;67:71–76.

Sterrett SG. Physical models and fundamental laws: using one piece of the world to tell about another. Mind Soc. 2002;3:51–66.

Sterrett SG. Models of machines and models of phenomena. Stud Philos Sci. 2006;20:69–80.

Thornton S. Karl Popper. In: Zalta EN, editor. The Stanford encyclopedia of philosophy (Summer 2009 ed). https://plato.stanford.edu/archives/sum2017/entries/popper/.

Tien C. Granular filtration of aerosols and hydrosols, series in Chemical engineering. Boston: Butterworths;1989.

Troelstra AS, Schwichtenberg H. Basic proof theory, 2nd ed. Cambridge: Cambridge University Press 2000.

Velasco J, Dobarganes C. Oxidative stability of virgin olive oil. Eur J Lipid Sci Techn. 2002;104:661–676.

Weber M. Objectivity' in social science and social policy (Translated and edited by Edward Shils, Henry Finch). Glencoe, Illinois: Free Press; 1949. pp. 49–112.

Wittgenstein L. Philosophical investigations. London: Blackwell; 1953.

Conclusions

**Antonios Kanavouras, Frank A. Coutelieris, Kostas Theologou
and Spyridon Stelios**

Through the definition of knowledge, one may conceive not only the core of knowledge but, accordingly, that of science itself, as well. But though scientific knowledge is linked through this tripartite 'umbilical cord' to knowledge, there is a significant difference between them. And this is the use of the laboratory as an epistemic know-how asset, where scientists seek answers about the world by looking at phenomena. Following Immanuel Kant's conceptual schema, one can assume that as senses capture in any form and origin data through the measuring instruments and devices, it allows the knowledge to develop only at phenomenal states; states that have, at any given time, a different spatiotemporal frame. Because of such a broadness and variability in their manifestation, understanding and interpretation of natural phenomena has always been a major challenge for scientists.

In this context, the combination of theoretical explanatory knowledge with mathematical modeling can fulfill the diachronic explanatory purposes of science. It is a new, sophisticated scientific language that is more accurately explaining and recording experimental results, bridging the gap between theoretical and practical research, and providing a methodological background (a road map) for organizing any future research activity. A close relationship between engineering and existing or produced knowledge about phenomena is obvious and this specific relationship is what this study attempts to represent via a mathematical model. More specifically, as shown in the second chapter, scientific representation could be, by means of artifacts, both concrete (graphs, scale models, monitor displays, etc.) and abstract (mathematical models, especially when infinity and/or infinitesimal play a role).

According to this, the model (or knowledge classification-transformation method) introduced here is not a scale model but an abstract, theoretical one; i.e. a theoretical construction that could be applied in a wide field of testing.

A specifically and comprehensively address of practical research and experiment by discussing the process of validation and verification of knowledge accomplished by the experimental method, has also been provided. This is a key element in the scientific formation of knowledge, as underlined in all chapters of this book. Lab

© The Author(s) 2018
F. A. Coutelieris and A. Kanavouras, *Experimentation Methodology
for Engineers*, SpringerBriefs in Continuum Mechanics,
https://doi.org/10.1007/978-3-319-72191-0_5

measurements correspond to data or other measures deriving from the artificial representation of natural phenomena under specific circumstances. This kind of representation does not downgrade nor diminishes the instrumental possibility they offer us in order to understand and explain the world in a rather accurate way. Yet their validity is optimal when grounded on the use of strict a priori concepts such as those materialized via mathematical modeling.

In this case, the model addresses in a comprehensive manner the problem of similarity among natural phenomena. What is particularly important is that this attempt takes place not only in terms of philosophy but also in terms of a mathematical treatment. It provides a revisit and review of natural sciences from the perspective of hermeneutic philosophy and in doing so it succeeds in mathematically capturing the spirit of the whole of science through the measuring of the individual, and vice versa.

Coming back to the notion of similarity one distinguishes two modes:

- The first mode is between measurements of phenomena and the model's form (i.e. scientific representation). What is the linkage between the model's mathematical terms and measurement procedures (for instance, C_A = concentration of substance A)? An obvious answer it the pre-existing knowledge about applied sciences and especially engineering. This provides the necessary background for the application of the theoretical model. This also reminds us that while coordination requires for its possibility some recognized empirical regularities, it is also required for the new theoretical assertions to have any empirical content at all. So, empirical measurements do play a role.
- The second mode is between phenomena. Which of them are eligible for induction into that specific methodological scheme? In this context, this engineering model could serve as an independent selection factor that sets the required scientific criteria. And that is the great innovation and originality behind its conception and formulation.

Any systems, formed and controlled by a number of physical phenomena allowing for the systems, possess many of their characteristic intrinsically coherent properties. For this work, the goal has been to approach the system holistically and reveal the cohesive points of the phenomena. Because of the intrinsic complexity and uncertainty that such a system implicitly presents, we needed to work on both an abstract as well as a penetrating way, at erudite and meticulous applications. Furthermore, the particular conditions covering as an "atmosphere" the systems are eventually controlling the progress of the phenomena. Such conditions need to be perceived and included as part of the system, fine-tuned at various levels during the experimental phase allowing via their impact on the system properties, the ably combination of the outcome data.

Within the goals of this work, the authors aimed in a solid knowledge based practice that has eventually derived via a straight forward mathematically confirmed procedure. Overall, the procedure proposed can evidently stand as a "technology" for that it can provide pragmatic effects in the experimentation set up. Moreover, the procedural use of the knowledge (input) classification (process) and the relevant treatments (transformations) produces realistic results (outcome) and the matrix itself is integral part of these results.

Accordingly, the book develops on the basis of the available knowledge treated in the view of, at least, providing the directions to gain new one, or set the potential actual conditions for that. Because the classification of knowledge may provide the so far, experimentation gaps and the non-applied conditions to fill them in, the whole proposal has also a direct advisory role and a potential application for filling the potential research gaps. Both shall derive from within the knowledge classification procedure itself, as this is solely based on experience of the physical phenomena on law base description, rather than subjective outcome, and the consequent cognition of natural systems.

A well-engineered system, can at the same time be a solid communication system among its structural constituents and the external environment. That system is actually a device, within which knowledge is not simply shelf-generated but it is created only when the correct, real, clear and justified information on the phenomena are feed backed internally and equally qualified information are to be received from the external environment. It is this process that keeps the science alive, via reducing potential mistakes, minimizing the stochastic disturbances that may cause damages beyond recuperation to its internal code ending up to the disruption of is functionality. Therefore, a safe conclusion may be derived on the validity of this approach regarding two major areas of interest, namely, the methodological approach and the consequent engineering tool.

The classification of existing knowledge regarding a specific phenomenon along with a disclaiming hypothesis, contains prerequisites of the description of the systemic phenomena to be reproduced for investigation, prerequisites for the identification of knowledge gaps through the recommended classification of the existing knowledge and guidelines to adequately design and perform the experimentation, in order to not simply to add fragments of knowledge in the field, but more over to embed the entity as a researcher himself in a dialectic relationship with the holistic and sustainable understanding of the world of phenomena in question.

The classification of the properties and the following mode of combining the principles of the system, allow for a low risk, shelf-assessing approach. In support to the value of this approach, the authors of this book have implemented a mathematically synthetic view, that possesses an objective validity free of the subjective and weaknesses. Such a view was highly confidently proposed as a methodological transition through different stages of a system's consideration. Such a transition could be sufficient and successful applied only when the methodology of matrix classification being presented herein, is apparently followed. More precisely, any accurate model definition must be the result of the efficient fulfillment of a 4×3 classification matrix, where each cell corresponds to a specific situation/parameter. This matrix actually includes all the possible categories and levels. Furthermore, the proposed process requires and imposes the necessity for the definition of rules and conditions under which it may be used. For that, the overall transition through the columns of the matrix could be adequately posed through three specific rules that are valid for controlling a system, when to be described in the way suggested here. This transition through the three columns follows the simple rule of dimensional increment: from the one-dimensional description (column 1) one may pass to a multi- but finite-dimensional space (column 2) to, finally, arrive in a vector space of infinite dimension (column 3). Likewise, the mathematical treatment followed the same trend: from a single equation (column 1) to a system of equations (column 2), while the last column 3 asked for a kind of different specific mathematical manipulations (asymptotic procedure, generalized integrations, etc.) in order to satisfy the transitions.

Additionally, knowledge classification may act in part, as a repository of information for modeling the potential disclaim of the system's hypothesis under increased risk. Categories, descriptors, classes and specifications are to be placed within this matrix, when designing experiments and selecting packaging materials, processes and means. Therefore, we would expect with high anticipation that this methodology will allow for constructive criticism among research groups, seeking for research accuracy, minimum research waste and deeper knowledge of the phenomena.

Furthermore, and based on the concept of similarity between physical phenomena, we present here a strict mathematical formulation for the world of physical phenomena of engineering interest. In specific, it is proven here that the set of all the perceptions of a phenomenon under investigation, sustained by closely defined operations, constitutes a four-dimensional vector space. These operations are quite similar to the conventional operations of summation and multiplication, equipped with all the necessary properties. For knowledge classification being essential for identifying similarities among perceptions of phenomena, a non-linear mapping over this vector space has been also defined to describe mathematically the similarity concept. Thus, the expressions of the necessary criteria for identifying similarity are simply translated to the estimation of the values for the crucial parameters affecting the problem. Finally, this mathematical treatment, not only allows for deep insight on a specific phenomenon, but also can identify the lack of knowledge, thus might direct the research on the field.

Closing Note

The consequent usage of the operational profile of a system, justified, among others, through the physical and mathematical theories, the cognition on optimum alternative solutions, benefits, risks, factorization and analysis of means and targets, will lead to a methodology concept for an engineered design of the systemic descriptors.

Considering every contemplative experience as an instrumental act in the further production of meanings, then ideas are the intellectual tools at the service of experimental procedures, conducted to resolve empirical problems. The more practical the idea, the closer the experience to achieving the overcoming of the limitations imposed by the phenomena. An additional mean towards decisive opening of the phenomena could be the technology and related engineering tactics. These are, combined, rationally selected methods, aiming at absolute efficiency in any field or type of experimental operations. Nevertheless, the human mind needs also to move along and ask the questions and definitions independently. Otherwise, science shall proceed with endless reproductions of the natural phenomena—superficially impossible to be completed ever—in order to reassure the conclusions on the true picture of the phenomena and their integration into nature.

In that sense, modern experimental science is in need as an art of control for eventual engineering of the systems, rather than an art of accepting things as they are. As a result, experimentation transforms the space—medley into space—system that be handled for quantifiably matched to mathematically separated measurements, imported into mathematical judgements as variables. A potential replacement of longstanding techniques by new ones, nevertheless obeying to comparatively different principles, allows the potential advancement of those research community members that wish to reaffirm the substantial rationality of scientific approach and results. The mathematization of natural phenomena simultaneously signifies the presentation, in an organized form, of all the theoretical proposals and results that we have been able to formulate through the empirical and aesthetically selected experiences. Through this organization, where each proposal arises from the previous one, fundamental properties of various phenomena can be explored and clarified methodically, on the basis of available knowledge.

In this way, the filtered simplification of the phenomenon as a whole and partly of the scientific field in which it belongs in with its problems, questions and knowledge, can be achieved via generic formulas. The simple development of such formulas may provide all the necessary equations requiring nothing more than algebraic calculations, on the basis of a regular and uniform step-by-step procedural concept, as presented herein in terms of the cycle of understanding. The coefficients of this transposition shall determine the revealing level and process for new ones, thus promoting innovative series of point-to-point operations. Such progressive and coordinated series of experiments exist within the classified knowledge satisfy the knowledge and understanding of our world.

Erratum to: Experimentation Methodology for Engineers

Erratum to:
F. A. Coutelieris and A. Kanavouras, *Experimentation*
*Methodology for Engineers***, SpringerBriefs in Continuum**
Mechanics, https://doi.org/10.1007/978-3-319-72191-0

In the original version of the book, the affiliations of "Theologou" and "Stelios" have been changed from "Department of Humanities, Social Sciences and Law, School of Applied Mathematics and Physics National Technical University of Athens Athens Greece" to "NATIONAL TECHNICAL UNIVERISTY OF ATHENS, Department of Humanities, Social Sciences and Law" in Frontmatter as well as in Chaps. 2, 3 and 5. The erratum book has been updated with the changes.

The updated online version of this book can be found at
https://doi.org/10.1007/978-3-319-72191-0

Index

A

Anomalies, 16, 17, 56
a posteriori, 9, 11–13, 51, 85, 86
Application(s), 16, 20, 22, 24, 26, 27, 35, 38, 48, 53, 54, 57, 60, 63, 68, 69, 73–76, 81, 83, 90, 91, 94, 95, 108, 109
Appropriation, 22
Approximation, 75, 99
a priori, 9, 11–13, 33, 36, 51, 73, 86, 87, 108
Aristotle, 12, 14, 16, 18, 61
Awareness, 7, 40, 42, 65

B

Bacon, Francis Sir, 21, 25
Belief, 5–7, 14, 15, 19, 37, 52, 61, 64
Berkeley, George, 11

C

Calculation, 62, 75, 99, 111
Causality, 13, 76
Circulation, 22
Classification, 39, 42, 46, 47, 55–57, 72, 81, 84, 85, 87, 88, 90, 94, 96, 101, 107, 109, 110
Coherence theory, 8
Complexity, 1, 34, 58, 59, 68, 71, 76, 83, 84, 88, 94, 108
Contingency, 13, 49, 85
Correspondence theory, 8
Criterion, 16, 17, 23, 41, 54
Cycle of understanding, 1, 44, 46, 56, 111
    hermeneutic, 41, 55, 56, 72

D

Data, 13–15, 18, 20–22
Deduction, 21, 44, 72
Dependence, 11, 13, 38, 44, 99
Device(s), 21, 27, 32, 34, 35, 41, 42, 45, 50, 56, 59, 62–64, 66–70, 72, 107, 109

E

Efficiency, 3, 45, 60, 68, 70, 90, 100, 111
    effectiveness, 68, 71
Einstein, Albert, 17
Empiricism, 2, 10, 18, 38, 39, 44, 49
Entity, 23, 56, 69, 109
Episteme (επιστήμη), 5
Epistemology, 5, 7, 9, 14
    epistemological, 7, 22, 38, 53, 59
Equipment, 3, 14, 21, 23, 35, 49, 52, 56, 57, 59, 62–67, 69, 70, 83
Error, 1, 6, 22, 25, 36, 54, 59, 64, 66, 72
Ethical values, 22
Evidence, 2, 3, 6, 7, 9, 11, 15, 18, 19, 23, 24, 37, 39, 47, 53, 54, 55, 84
Existence, 1, 2, 9, 11, 13, 18, 24, 35, 36, 75
    non-existence, 13
Experience, 1–3, 9, 11–13, 15, 16, 22, 23, 35–37, 40, 41, 45, 47, 48, 51, 52, 54, 57–59, 64, 65, 67, 70, 85–88, 90–92, 109, 110
Experiment, 10, 20, 21, 24, 27, 32, 33, 46, 49, 50, 57, 60–63, 65, 66, 68–70, 75, 102, 107, 110, 111
    experimentation, 2, 3, 21, 27, 33, 35, 41, 49, 50, 56, 57, 59, 61–65, 67–71, 75, 84, 86, 108, 109, 111
    experimenter, 34, 45–47, 49, 62, 64, 63, 65, 67–70, 72
Explanation, 15, 22, 33, 38, 40, 50, 53, 54, 84

F

Fact, 2, 6, 9, 10, 12, 15–19, 23, 60, 61, 65, 71, 74, 84, 86, 94, 102
Falsification, 15, 16, 18, 71
Feyerabend, Paul, 16, 18, 24

G

Gadamer, Hans Georg, 40

© The Author(s) 2018
F. A. Coutelieris and A. Kanavouras, *Experimentation Methodology for Engineers*, SpringerBriefs in Continuum Mechanics, https://doi.org/10.1007/978-3-319-72191-0

Gärdenfors, Peter, 7
Gettier, Edmund L., 6
Gödel, Kurt, 36

H
Habermas, Jürgen, 55
Hacking, Ian, 18, 21, 24, 62
Helicocentrism, 18
Human factor, 34, 36, 43, 57, 69
Hypothesis, 2, 3, 19, 20, 27, 31–33, 35, 42,
    45–49, 52, 53, 58, 60–62, 68, 70, 71, 82,
    84, 85, 87–91, 109, 110
  hypothetico-deductive method, 53, 54

I
Idea platonic, 37
Induction, 14, 15, 21, 108
Information, 7, 10, 33, 34, 37–39, 41, 46, 49,
    51, 52, 54, 56–58, 65, 100, 109, 110
Institution, 22
Instrument, 16, 21, 23, 24, 55, 59, 69, 70, 107
  instrumentalism(ist), 23, 82
Interpretation, 8, 25, 32–34, 39, 45, 52–54, 71,
    72, 82–84, 88, 107

J
Judgment, 11–14, 34, 51, 54, 56, 73, 85–88, 91
Justification, 6, 7, 20, 33, 41, 83, 87

K
Kant, Immanuel, 9, 11–14, 76, 107
Knowledge, 1–3, 5–17, 20–27, 32–34, 37–40,
    44–49, 51, 52, 54–58, 61, 63, 65, 66, 70,
    72, 73, 75, 76, 81, 84–88, 90–94, 102,
    107–111
  classification scheme, 37, 54, 65, 87
  cognition, 1, 11, 12, 21, 42, 45, 46, 61, 66,
    73, 87
  new knowledge, 9, 22, 25, 31, 39, 45, 92
  normative knowledge, 5, 35, 41
  practical knowledge, 5, 32, 33, 37, 40, 49,
    63, 76, 86, 90, 107
  scientia, 5
  scientific knowledge, 2, 11, 13, 14, 16, 34,
    37, 42, 54, 55, 84, 107
  similarity of, 37, 42
  situational knowledge, 22
  theoretical knowledge, 5
Koutoungos, Aristophanes, 7
Kuhn, Thomas, 15, 16, 18, 24, 32, 45, 83, 84

L
Lakatos, Imre, 16, 17
Lavoisier, Antoine, 16

Law scientific, 10, 20, 25, 34, 38, 57
Limitation, 75, 110

M
Mathematics, 9, 11, 24–26, 36, 51, 73–76, 87,
    88, 92
MATRIX, 3, 95
  the matrix scheme, 94
Matter, 21, 47, 48, 50, 65, 67–69, 72, 83, 88,
    89, 91, 93, 95, 96, 100, 101
Maxwell, James Clerk, 16
Measurement, 18, 21, 24, 26, 56, 63, 66, 67,
    82, 83, 108
Mechanics, 17, 60
Method, 6, 10, 15, 16
Modality, 13
Model, 10, 23–26, 32, 37, 48, 61, 62, 64,
    74–76, 83, 86–88, 90, 99, 100, 107–109
Modeling, 10, 23, 24, 32, 74, 83, 107, 108, 110
  stochastic, 74, 109
Monism, 18
  methodological monism, 18

N
Nature, 2, 7, 8, 10, 17, 20, 23, 26, 27, 37, 38,
    56, 61, 63, 74–76, 83
Necessity, 11, 13, 51, 72, 76, 85–88
Negation, 6, 12
Newton, Isaac, 17, 21, 32, 91
Nooumenon-a, 13
Nozick, Robert, 5

O
Observer, 2, 35, 63
Outcome, 1, 2, 6, 23, 32, 33, 35, 44, 47, 48, 50,
    57, 60, 61, 65, 68, 70, 71, 85, 87–89, 91,
    93, 94, 97, 99–101, 109

P
Paradigm, 1, 16, 18, 20, 33, 34, 39
Pascal, Blaise, 5
Perception, 2, 21, 27, 32, 34, 35, 42, 58, 61, 84,
    86, 87, 91–94, 99, 110
Perspective, 10, 24, 37, 39, 54, 82, 108
Phenomenon-a, 3, 32
  physical phenomenon, 42, 43, 81, 93
Philosophy of science, 9–11, 15, 16, 18, 24
Physics, 9, 13, 18, 26, 32, 50, 60, 71, 73, 76,
    83, 90, 102
Plato, 5, 50
Plurality, 12, 53
Popper, Karl, 14–17, 33, 37, 38, 41, 44, 53, 54,
    59, 71, 72, 76, 83, 93
Popularization, 22

Possibility, 11, 20, 31, 41, 49, 57, 71, 83, 91, 108
    impossibility, 13
Practice, 16, 17, 19, 37, 38, 69, 88, 90, 108
Prediction, 15, 23–25, 27, 32, 43, 64, 85
    predictive, 10, 31, 43, 83, 86, 99
Prigogine, Ilya, 20, 31, 32
Process, 1, 2, 6, 7, 14, 16, 20–22, 32, 33, 35, 37, 39–41, 44, 45, 47, 51, 52, 54, 57, 58, 60, 62, 69, 71, 73, 88, 109
Properties, 3, 15, 25, 47–50, 84
Psillos, Stathis, 15
Ptolemy, 16

Q

Quality(ies), 21, 22, 47, 48
Quantity(ies), 12, 33, 44, 69, 74, 75, 84, 89

R

Reality, 11, 25, 34, 63, 75, 88
    realism, 11
    realist, 23, 37, 109
Relationship(s), 2, 3, 7, 8, 10, 15, 27, 33, 37, 46–48, 50, 55, 57, 59, 60, 62, 66, 67, 84, 89, 91, 93, 98, 100
Relativity theory, 17
Representation, 11, 12, 23, 25, 52, 87, 108
Research program, 17
Result(s), 9, 15, 24, 31, 38, 39, 42, 46, 52, 58, 60, 61, 65, 67, 71, 82, 83, 85, 91, 99, 102, 107, 111
Rule(s), 7, 8, 18, 21, 32, 58, 69, 71, 87, 90, 94, 109
Russell, Bertrand, 12

S

Salmon, Merrilee H., 16, 25
Science, 9, 13–19, 36, 51, 59, 75, 108, 109, 111
    pseudoscience, 1, 17
Simulation, 44, 75

Social constructivism, 22
Speculation, 19, 62, 75
Structure, 1, 10, 14, 15, 25, 37–39, 50, 54, 59, 60, 64, 68, 72, 75, 86, 93
System, 2, 18, 24, 26, 34, 36, 37, 41, 42, 45, 46, 54, 55, 59, 65, 67, 68, 71, 72, 75, 81, 83, 84

T

Technology, 10
    technological, 1, 10, 20, 21, 31, 32, 35, 44, 45, 55, 63, 83
Theaetetus, 5
Theory, 6, 8, 19, 15–26, 32, 38, 43–45, 50, 62, 63, 70, 76, 94
    meta-theory, 1
Totality, 12, 41, 69, 70
Truth
    truth value, 8

U

Uncertainty, 1, 7, 27, 108
Unity, 12, 75, 87
Universality, 5, 51, 85

V

Validation
    validity, 55, 60, 66, 82, 108, 109
Value, 2, 5, 8, 22, 41, 42, 61, 66, 74, 91, 100
    boundary value, 24, 40, 97, 99–101
Van Fraasen, Bas, 15, 21, 26
Vernunft, 12
Verstand, 12
Vorgriff (anticipation), 54
Vorhabe (intention, plan), 54
Vorsicht (caution), 54

W

World, 2, 8, 12–14, 21, 39, 70, 108
    physical world, 2, 16, 37, 46, 74, 84

Printed in the United States
By Bookmasters